The Hunter-Gatherer Principle

VINCIT
OMNIA
VERITAS
I.S.

✒ LOCHLAINN SEABROOK WRITES IN THE FOLLOWING GENRES ✒

Alternate History
American Civil War
American History
American Politics
American South
American West
Anatomy and Physiology
Ancient History
Anthologies
Anthropology
Apocrypha
Aquariology
Archaeology
Astronomy
Aviation
Aviation History
Behavioral Science
Biblical Exegesis
Biblical Hermeneutics
Bioarchaeology
Biography
Children's Books
Christian Mysticism
Clinical Studies
Coffee Table Books
Coloring Books
Comparative History
Comparative Mythology
Comparative Religion
Constitutional Studies
Cooking
Cultural Anthropology
Cultural History
Cultural Studies
Cryptozoology
Diet and Nutrition
Earth Sciences
Ecology
Ecotourism
Education
Encyclopediography
Entertainment
Environmental History
Environmental Science
Epistemology
Ethnobotany
Ethnology
Ethology
Ethnomusicology
Ethnic Studies
Etymology
European History
Evolutionary Anthropology
Evolutionary Biology
Evolutionary History
Evolutionary Psychology
Exobiology

Exposés
Family Histories
Film
Folklore
Genealogy
General Audience
Geography
Geology
Genetics
Ghost Stories
Gospels
Health and Fitness
Historical Ecology
Historical Fiction
Historical Nonfiction
Historiography
History of Medicine
History of Science
Hobbies and Crafts
Human Evolution
Humanities
Humor
Ichthyology
Illustrations
Inspirational
Interviews
Journalism
Law of Attraction
Lexicography
Life After Death
Lifestyle
Literature
Marine Biology
Matriarchy
Medical History
Memoir
Men's Studies
Metahistory
Metaphysics
Military
Military History
Mysteries and Enigmas
Mysticism
Mythology
Natural Health
Natural History
Natural Philosophy
Natural Science
Oceanography
Onomastics
Paleoanthropology
Paleoecology
Paleography
Paleoichthyology
Paleontology
Paleozoology
Paranormal

Parapsychology
Patriarchy
Performing Arts
Philosophy
Philosophy of Science
Photography
Physical Anthropology
Pictorial
Poetry
Politics
Prehistoric Life
Prehistory
Presidential History
Primatology
Primary Documents
Prophecy
Psychology
Quiz
Quotations
Recollections
Reference
Religion
Revolutionary Period
Science
Scripture
Self-help
Social Sciences
Sociology
Southern Culture
Southern Heritage
Southern Narratives
Southern Traditions
Speeches
Spirituality
Spiritualism
Sport Science
Symbolism
Technology
Thanatology
Thealogy
Theology
Theosophy
Travel
UFOlogy
Vexillology
Victorian Era Studies
Victorian Medicine
Visual Arts
War
Western Civilization
Wildlife
Wildlife Photography
Women's Studies
World History
Writing
Young Adult
Zoology

Mr. Seabrook does not author books for fame and glory, but for the love of writing and sharing his knowledge.

Be curious, not judgmental.

✒ SeaRavenPress.com ✒

Warning: SEA RAVEN PRESS BOOKS WILL EXPAND YOUR ★ MIND!

THE HUNTER-GATHERER
PRINCIPLE

EVOLUTIONARY BIOLOGY & THE CASE FOR SEX-BASED FEMALE SPORTS

LOCHLAINN SEABROOK

Best-Selling Author, Award-Winning Historian, Acclaimed Artist

Diligently Researched and Generously Illustrated
by the Author for the Elucidation of the Reader

2025

Sea Raven Press, Park County, Wyoming, USA

THE HUNTER-GATHERER PRINCIPLE

Published by
Sea Raven Press, LLC, Founded 1995
Cassidy Ravensdale, President
Park County, Wyoming, USA
SeaRavenPress.com

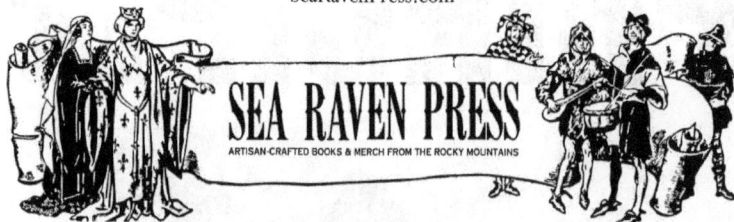

SEA RAVEN PRESS
ARTISAN-CRAFTED BOOKS & MERCH FROM THE ROCKY MOUNTAINS

PRINTING HISTORY
1ˢᵗ SRP paperback edition, 1ˢᵗ printing, May 2025 • ISBN: 978-1-955351-56-0
1ˢᵗ SRP hardcover edition, 1ˢᵗ printing, May 2025 • ISBN: 978-1-955351-57-7

ISBN: 978-1-955351-56-0 (PAPERBACK)
Library of Congress Control Number: 2025937597

The Hunter-Gatherer Principle: Evolutionary Biology and the Case for Sex-Based Female Sports, by Lochlainn Seabrook. Includes an introduction, illustrations, index, endnotes, appendices, and bibliography.

ARTWORK
Front and back cover design and art, book design, layout, font selection, and interior art by Lochlainn Seabrook.
All images, image captions, graphic design, and graphic art copyright © Lochlainn Seabrook.
All images selected, placed, manipulated, cleaned, colored, tinted, and/or created by Lochlainn Seabrook.
Cover photo: "Three Pro Female Athletes," copyright © Lochlainn Seabrook.

All persons who approve of the authority and principles of Colonel Lochlainn Seabrook's literary work, and realize its benefits as a means of reeducating the world about facts left out of mainstream books, are hereby requested to avidly recommend his titles to others and to vigorously cooperate in extending their reach, scope, and influence around the globe.

The views documented in this book concerning evolutionary biology and sex-based women's sports are those of the publisher.

WRITTEN, DESIGNED, PUBLISHED, PRINTED, & MANUFACTURED IN THE UNITED STATES OF AMERICA

REAL SCIENCE MATTERS

Dedication

To all biological female athletes.

Women polo players, 1920s.

Epigraph

"Girls need safe, healthy and supportive environments to grow and excel. Policies need to tap the power and potential of physical activity and sport to advance girls' health, physical and emotional development, social well-being and educational aspirations and achievements. Efforts must be directed toward increasing girls' participation in physical activity and sport."

From the President's Council on
Physical Fitness and Sports Report, 1997

Compiled under the direction of the
Center for Research on Girls and Women in Sport

Girls on a hike, 1920s.

CONTENTS

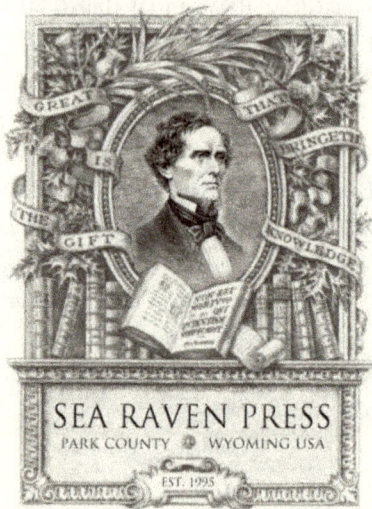

SEA RAVEN PRESS
PARK COUNTY ∞ WYOMING USA
EST. 1995

"Books invite all; they constrain none."
Hartley Burr Alexander (1873-1939)

NOTES TO THE READER

---◆●◆---

SCIENCE AND THE LAW
TRUMP FEELINGS AND EMOTIONS

THIS IS NOT A POLITICAL BOOK. Though some would try to make it so. It is a science book. More importantly, I am not anti-transgender and this book is not anti-transgender. I have no personal feelings for or against transmen and transwomen, and I support their right to identify as whichever or whatever gender they like. Indeed, I have libertarian views on most social issues, including transgenderism, and believe that an individual should able to pursue happiness in whatever manner they desire: *as long as it does not violate the constitutional rights of others*.

When it comes to athletic competition, however, the reality is that men identifying as women in order to compete in women's sports is a clear violation of both women's innate civil rights and Title IX, the latter which is an elaboration of the 14th Amendment whose Equal Protection Clause, in turn, guarantees equal protection under the law to *all* citizens. Allowing a 6 foot, 200 pound man to play with or against a five and a half foot, 125 pound woman is *never* fair and it is *not* protecting women. It does the opposite: it jeopardizes their health, well-being, and even their lives. As a blatant violation of Title IX, not only is this the very definition of "unsportsmanlike behavior," it is, in fact, a crime under current law.

Besides being demeaning and plainly dishonest, it is also unjust and dishonorable to women as well as an infraction of Federal law. This includes aspects of both the 5th Amendment and the proposed 28th Amendment—men playing in women's sports also creates ethical and moral issues, and perhaps most significantly of all, safety issues.

For these reasons, and others that I will discuss in the following pages, I believe that biological men, however they self-identify, should and must be permanently banned from all federally-funded women's sports. For, as my British cousins have recently legally established: "A woman is defined by biological sex," not by gender preference.

DEFINITIONS
To make this perfectly clear and avoid confusion throughout the following pages, it is important that we establish several important definitions: What is a man, a woman, a transman, and a transwoman?

THE TWO BIOLOGICAL *SEXES* DEFINED
A biological male (boy/man) is an individual born with one X chromosome and one Y chromosome. A biological female (girl/woman) is an individual born with two X chromosomes.

THE TWO TRANS *GENDERS* DEFINED
An individual born a male (boy/man), that is, with one X chromosome and one Y chromosome, but who identifies as or feels like a woman, is a transgender woman. An individual born a female (girl/woman), that is with two X chromosomes, but who identifies as or feels like a man, is a transgender man.

WESTERN CIVILIZATION UNDER THREAT
For the record, and as this book will make implicitly clear, a transwoman is not a woman any more than a transman is a man. These are metaphysical terms, deviously invented by Left-wing activists to disrupt, reorganize, and force Western society down a path toward socialism and eventually communism. By accepting the insertion of self-identifying transgender policies, opinions, and ideologies into modern culture, we inch ever closer to the downfall and ultimate ruination of all that our ancestors began constructing

Women canoers, Custer State Park, South Dakota, 1920s.

in ancient Greece (and later Rome) some 2,700 years ago: Western civilization. Will we stand aside and watch idly as our very identity is devalued, diluted, and ultimately extirpated?

For the record, according to the biological sciences, however one chooses to personally view this issue, a transwoman is still a man and a transman is still a woman. Nothing can or ever will change these facts.

L.S.

INTRODUCTION

SCIENTIFIC TRUTH

S HOULD MEN IDENTIFYING AS WOMEN—or, depending on your point of view, men claiming or pretending to be women—be allowed to participate in female sports, either on the same teams or on opposing teams? And by the same token should these same biological men posing as females be allowed to use women's locker rooms, dressing rooms, and other female-only spaces, while taking and winning honors, titles, and awards that would normally go to biological women? Should transgenders, who represent a mere 0.6 percent of the American population,[1] be allowed to override the desires, traditions, customs, and beliefs of the other 99.4 percent?

There are many *subjective* ways to answer these questions. We all have and are entitled to our personal opinions and feelings, after all.

Nonetheless, there is only one *objective* way to answer the above questions, and that answer has been unequivocally obvious for at least 1.5 million years. And the answer remains the same whether you are a theist or an atheist, a Conservative or a Liberal, a layperson or a professional scientist, a heterosexual or a homosexual, cisgender or transgender, intersexual or bisexual, an anarchist or a communist. In all instances the answer is a resounding "no." Biological human males, that is, individuals assigned to the male sex at birth because they possess an X and a Y chromosome, should not be allowed to play alongside or compete against biological human females. There are a host of reasons for this, including scientific reasons, ethical reasons, moral reasons, and safety reasons.

Female rock climber, Mount Desert Island, Maine, 1920s.

In the following pages I will discuss the biology underlying this statement, reinforced with copious evidence that will convince all but the most close-minded alethophobes.

All others, however, will instantaneously see the truth for what it is: an uncompromising biological reality that overshadows all personal views and all political ideologies. This is why, after all, it is called truth. More specifically, in our case it is *scientific truth* that we will be examining and discussing.

Lochlainn Seabrook
Rocky Mountains, USA
May 2025

Miss Dorothy D. Smith, five times National Archery Champion, 1920s.

CHAPTER ONE

THE BIOLOGICAL REALITY OF BEING HUMAN

WE LIVE IN A SOCIALLY confused day and age in which blatant scientific illiteracy and stunning willful ignorance are commonplace, even accepted as the norm in some circles. It is a time period in which the ill-informed can arrogantly and publicly claim that there are "hundreds of different genders," and in which radical Left-wing activists cunningly impersonate judges, educators, politicians, journalists, and news anchors.

Women campers at a state park, 1920s.

In such a world where up is down and down is up, it is not enough to simply state that there are actually only two sexes, that there always have been and always will only be two sexes; that someone born with an X and a Y chromosome is a biological male and that someone born with two X chromosomes is a biological female; that, in essence, a man is a man and a woman is a woman; that a transwoman is still a man, and that a transman is still a woman, and so on.

Fundamental biology has decreed that the gender binary (there are only two sexes) is real, for it is prehistorically and genetically encoded, and therefore immutable. As such, realists must meet the blatant nonsense that attempts to deny and counter it head on using logic and well established science.

PHYSICAL SUPERIORITY OF THE HUMAN MALE

We must begin, of course, with the most obvious physical fact pertaining to *Homo sapiens*: *The average man is physically superior to the average woman*. This is why, looking more closely at this fact, we find that in nearly every *physical* (though not cerebral) sport, men easily outperform women. This is not opinion, sexism, misogyny, or even androphilia. Let us look at the evidence. As of early 2025, for example:

- The current men's 26 mile marathon world record is 2:00:35.
- The current women's 26 mile marathon world record is 2:09:56.

- The current men's long jump world record is 8.95 meters.
- The current women's long jump world record is 7.52 meters.

- The current men's high jump world record is 2.45 meters.
- The current women's high jump world record is 2.10 meters.

- The current men's 100 meter sprint world record is 9:58.
- The current women's 100 meter sprint world record is 10:49.

- The current men's 200 meter sprint world record is 19:19.
- The current women's 200 meter sprint world record is 21:34.

- The current men's 500 meter middle-distance run world record is 57:69.
- The current women's 500 meter middle-distance run world record is 1:05.63.

- The current men's 1500 meter middle-distance run world record is 3:27.65.
- The current women's 1500 meter middle-distance run world record is 3:49.04.

Female golfing pro, 1920s.

- The current men's 3000 meter middle-distance run world record is 7:17.55.
- The current women's 3000 meter middle-distance run world record is 8:06.11.

- The current men's 5000 meter long-distance run world record is 12:35.36.
- The current women's 5000 meter long-distance run world record is 14:00.21.

- The current men's 10,000 meter long-distance run world record is 26:11.00.
- The current women's 10,000 meter long-distance run world record is 28:54.14.

- The current men's 3000 meter steeplechase world record is 7:52.11.
- The current women's 3000 meter steeplechase world record is 8:44.32.

- The current men's freediving world record is 121 meters.
- The current women's freediving world record is 109 meters.

- The current men's shot put world record is 23.56 meters.
- The current women's shot put world record is 22.63 meters.

- The current men's javelin throw world record is 98.48 meters.
- The current women's javelin throw world record is 72.28 meters.

- The current men's sport climbing world record is 4.74.
- The current women's sport climbing world record is 6.06.

Women polo players, 1920s.

All-women's field hockey team, Yorkshire, England, early 20th Century.

- The current men's speed cycling world record is 16.42 mph.
- The current women's speed cycling world record is 13.23 mph.

- The current men's 50 meter freestyle swimming world record is 21.57.
- The current women's 50 meter freestyle swimming world record is 24.03.

- The current men's 100 meter freestyle swimming world record is 46.40.
- The current women's 100 meter freestyle swimming world record is 51.96.

- The current men's power lifting deadlift world record is 1,107.8 lbs.
- The current women's power lifting deadlift world record is 700 lbs.

- The current men's weight lifting clean and jerk world record is 589 lbs.
- The current women's weight lifting clean and jerk world record is 354.944 lbs.

- The current open men's 500 meter indoor rowing world record is 5:35.8.
- The current open women's 500 meter indoor rowing world record is 6:21.1.

- The current men's downhill speed skiing world record is 158.760 mph.
- The current women's downhill speed skiing world record is 153.530 mph.

- The current men's pole vault world record is 20 ft. 6.5 inches.
- The current women's pole vault world record is 16 ft. 7 inches.

- The current men's 1500 meter speed skating world record is 1:40.17.
- The current women's 1500 meter speed skating world record is 1:53.28.

- The current men's 5000 meter speed skating world record is 6:01.56.
- The current women's 5000 meter speed skating world record is 6:43.51.

- The current men's 400 meter hurdles world record is 45.94.
- The current women's 400 meter hurdles world record is 50.37.

And so on. Again, this is not sexism. It is statistical reality.

It is admittedly true that the strongest professional female power lifter would probably outlift the weakest amateur male. However, this same female could never outlift the strongest male power lifter. This is the difference; a difference that is readily recognized by every reasonable, rational, thinking human being.

SPORTS MAKE ADJUSTMENTS FOR THE FEMALE PHYSIQUE
This is why in nearly every sport where the foundation is primarily physical (as opposed to more intellectual sports like chess, backgammon, video games, table tennis, competitive ballroom dancing, etc., where men and women largely perform equally),[2] *biological males win out in all categories.*

This fact is one of the primary reasons why a 2025 poll found that 94 percent of Republicans and 67 percent of Democrats believe that transgender women, that is, biological males, should not be allowed to compete against biological women.[3] And it is why, on February 5, 2025, America's current president, Donald J. Trump, signed an executive order banning men from participating in women's sports. In defense of biological realism, our 47th chief executive rightly said of this radical Left-wing ideology:

> "It is demeaning, unfair, and dangerous to women and girls, and denies women and girls the equal opportunity to participate and excel in competitive sports."[4]

Mrs. Delphine Dodge Cromwell, the first woman speed boat race driver to compete in the exclusive Gold Cup regatta, Manhasset Bay, Long Island, New York.

This has been so obvious for so long that a number of sports transparently compensate for our built-in biological sex differences by altering or even lowering the standards for women—an effort to literally "even the playing field."

In high diving world championships, for instance, we find that the men's dive platform is 27 meters high, the women's is 20 meters high, an overt adjustment to the fact that women, *on average*, are physically smaller, less muscular, and anatomically more fragile than men. In the sport of discus throw we encounter the same sex-typed counterbalancing: Officials have decreased both the weight and the diameter of the women's discus: The men's discus weighs 2 kilograms and is 22 centimeters in diameter; the women's discus weighs 1 kilogram and is 18 centimeters in diameter.[5]

It is not just the mechanics of sports that are adjusted to fit the inherent physical inferiority of human females. Take boxing, for example. Men's championship fights are 12 rounds, three minutes each; women's fights, however, are usually 10 rounds, two minutes each. Experts claim that this disparity is *not* because "women are the weaker sex," but because there is a safety issue pertaining to women that is not as much of a concern with male boxers: Women's skulls are smaller, thinner, and more delicate, and so are more prone to concussion. Thus, cutting the time girls and women are in the ring is an attempt to reduce female brain injuries. While imposing this regulation on women boxers is medically correct, it does not change the fact that this rule has its origins

The "Discus Thrower" (*Discobolus*), by the Greek sculptor Myron, circa 450 B.C.

in biology. Again, the average human female is physically inferior to the average human male.

Then there is tennis: Men generally play five sets, women generally play three sets. The reason? Tennis organizations have agreed, since 1901, that women have neither the stamina nor the endurance to play five sets. While it is obvious that most female tennis players—being highly trained and skilled athletes—*could* actually play five sets, it is almost universally believed that due to fatigue setting in by the fourth or fifth set, both the live audience and the TV viewing audience would begin to lose interest, causing a corresponding loss in ticket sales, league profits, and ad revenue.

We also have the sport of rock climbing. Here again, due to obvious biological differences (which progressives misrepresent using the phrase "perceived differences") between males and females, not only are the two sexes divided into separate men's and women's categories, women are given less difficult routes than men.[6]

Why do so many physical sports make adjustments for women if there are "*no differences between men and women*," if "men and women are equal in all ways," and if "a woman can do anything a man can do," as we are repeatedly told by those who place feelings over facts?

How do we explain these physical differences? Surely these are not accidental. For God/Nature is not only purposeful in all its designs, it has rooted all our physical and mental structures and capacities in physicality, a physical law we call biology. As American biologist E. O. Wilson once stated, "there is a biological basis for all human behavior."[7]

To find the answer to our questions we must turn to the hard sciences, beginning with the Nature versus Nurture phenomenon, which is in turn associated with our separate male and female hardwired biological programming (note: this is quite different from our learned behavior, which is softwired). It is here that we find the concrete answer to the entire issue, with its many current questions, theories, opinions, and disagreements.

THE HUNTER-GATHERER PRINCIPLE

Thanks to our earliest known ancestors, *Homo habilis* and *Homo erectus*, whose skeletal structures suggest hunting and whose stone cutting tools suggest meat-eating—and in turn, a sex-based division of labor between males and female—we ourselves, modern *Homo sapiens*, clearly evolved from out of a hunter-gatherer milieu.[8] In other words, we inherited the bodies and minds of hunter-gatherers, making us in every way, in fact, 21st-Century hunter-gatherers.[9]

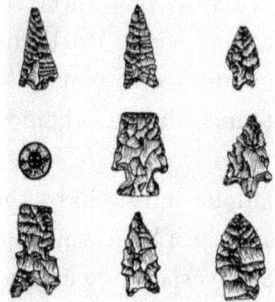

Stone projectile points and other physical evidence of the prehistoric hunting and gathering lifestyle.

From a purely biological-behavioral standpoint then, human males were evolutionarily programmed to be big-game hunters and warriors (protectors)[10] and to form small androarchies[11] ("brotherhoods," "bachelorhoods," or "fatherhoods") with other men. Human females, on the other hand, were evolutionarily programmed to be maternal caregivers (mothers) and food gatherers, and to form small gynoarchies[12] ("sisterhoods," "maidenhoods," or "motherhoods") with other women.[13]

While there is plenty of anecdotal *social* evidence substantiating this view, it is too easily debated to be considered irrefutably scientific. However, when we turn to *physical* evidence, or what is known as *sexual dimorphism* (the physical dissimilarities between males and females), a very different and very persuasive picture emerges.[14]

American hunter-gatherers: Crow and Shoshone men driving buffalo in what is now Lamar Valley, Yellowstone National Park, Wyoming.

Though progressive scientists like to downplay it, sexual dimorphism, which begins in humans at the age of one month and continues to develop through puberty and into early adulthood,[15] has had, and continues to have, an enormously profound affect on our species. Even more so when we look at what is called *male-biased sexual size dimorphism* (SSD) in our species, a closely associated topic of this book.

On average, men—*built on an archetypal big-game hunter design in which spatial skills, physical strength, prowess, teamwork, inventiveness, and endurance are key*—have much larger, wider, heavier, and thicker bodies, brains (brain dimorphism), skulls, jaws, bones, muscles, lungs, organs, blood vessels, hearts, canine teeth,[16] and even higher hemoglobin levels, than women. Additionally, males have lower voices, conspicuous facial and body hair, and different fat distribution than females (in men centered on the abdomen, in women centered on the hips and thighs). As these traits are inherent to the human male, this is why merely reducing testosterone levels in transwomen does not benefit biological female athletes. You cannot undo what Nature has wrought.

THE HUMAN PELVIS

The most noticeable physical difference between human males and human females (both internally and externally) is the pelvis, with the former being thicker, heavier, denser, deeper, and narrower, the latter being thinner, lighter, finer, shallower, and wider. These differences include blatant male and female variations of the iliac crest, acetabula, pelvic outlet, subpubic angle, obturator foramen, sacrum, ischial tuberosities, pelvic inlet, coccyx, greater sciatic notch, and ischiopubic ramus (subpubic concavity). Some parts that form the female pelvis, like the ventral arc, are completely lacking in the male pelvis.[17]

The two pelvises are so dissimilar in size, construction, shape, weight, and appearance that scientists, from paleoanthropologists and forensic detectives to medical doctors and archaeologists, can almost immediately distinguish between a male skeleton and a female skeleton—*even if it is millions of years old.* As the Hunter-Gatherer Principle reveals once again, the more robust male pelvis offers advantages for hunting and defensive actions (running, kicking, etc.), while the more gracile female pelvis offers advantages for birthing and carrying infants.[18] So striking are the highly contrasting sexual distinctions between human males and human females, it is little wonder that when subjectively rated, studies show that we humans rank as the eighth most visually sexually dimorphic primate out of 124 other primate species.[19]

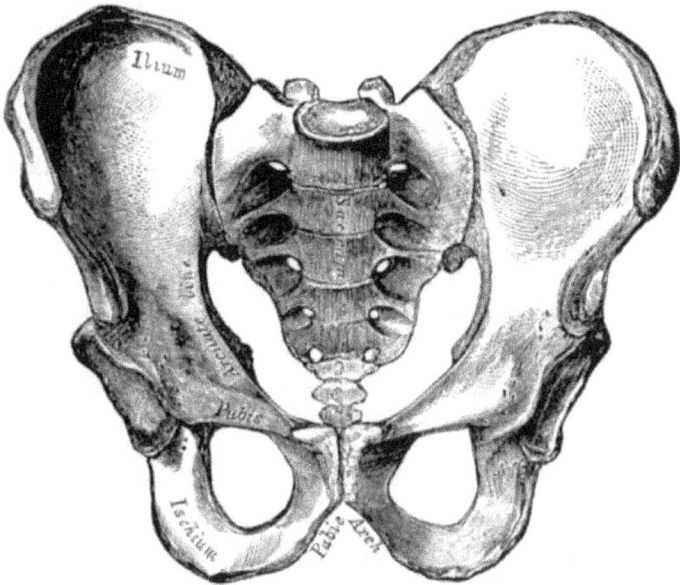

Average adult human male pelvis.

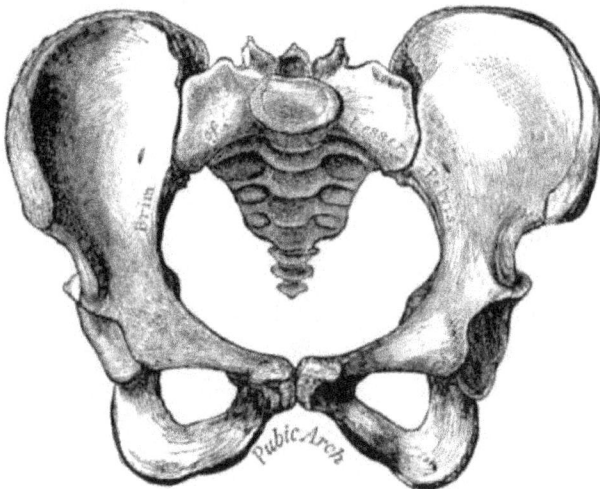

Average adult human female pelvis.

THE HUMAN SKULL

The second most physically dimorphic structure in the human body is the skull, with, as we have seen, the male's being larger and thicker, the female's being smaller and thinner. The diagrams below illustrate these differences.

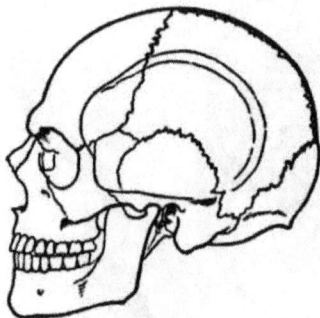

Average adult human male cranium.   Average adult human female cranium.

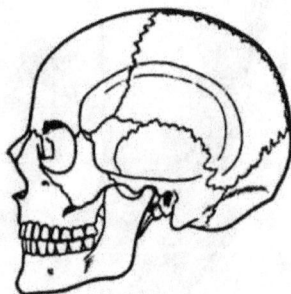

Even from a cursory glance one can easily see the advantages the male skull offers when it comes to hunting and protection: A smooth sloping forehead that deflects impacts; heavy protruding suborbital ridges (brow ridges) that help protect the eyes; and a massive lower jaw capable of receiving powerful blows. Overarching all of this masculine defensive osteology is the male calvaria (skull cap), comprised of dense bone that helps protects his all-important biological computer: the brain. When necessary, the human male's large, strong, and aggressive masculine facial (and bodily) appearance is very well suited to intimidating both animals and fellow humans.

The human female, on the other hand, hardwired for maternal and food-gathering concerns, would have received no benefits from possessing massive a skull like her male counterpart (in fact, it would have hindered her), and so evolved a smaller more delicate cranium, giving her a softer, gentler, more feminine neotenous appearance that is perfectly suited to child-rearing.

Placing these many stark male and female differences side by side, we are left with an inevitable biological result: Men, again on average, can run faster and further than women, lift heavier objects, jump higher and further, throw objects further, hold their breath longer, swim faster and further, dive deeper, climb faster, and overall withstand greater physical stress and discomfort over a longer period of time than women. This is why, as just one of many examples that could be given, scientific studies find that "the occasions where women were able to beat men in long-distance running events were very rare exceptions and happened only at recreational competitions, but never at professional competitions."[20]

THE SAFETY ISSUE

More to the point, it is also why competitive athletics and games, such as those performed at the Olympics, have always been traditionally gender binary, that is, separated into men's divisions and women's divisions: It would give men an unfair advantage to pit the two sexes together on, for example, a basketball court, a rugby pitch, a javelin field, a football field, or in a boxing ring. It would create unbalanced competitions, wholly favoring men.

Those female athletes who ignore this reality (or in many cases, those who are *forced* to accept this reality against their will)—that is, who have competed against male athletes, have often come away with serious even life-threatening physical and psychological damage. One such woman was Payton McNabb: During a volleyball match in 2022, the former North Carolina high school student suffered long-term physical and mental injuries caused by a biological male who identified as "transgender."[21]

The resulting serious posttraumatic effects from allowing biological males to participate in women's sports is widely known. U.S. Attorney General Pam Bondi recently said the following:

> "I've had the opportunity to speak to countless girls . . .
> across the country. . . countless parents, coaches, who all

virtually say the same thing: Why wouldn't anyone protect my athlete? Why wouldn't anyone protect my daughter? Why wouldn't anyone protect me? I've talked to girls who have lost roster spots and podium finishes. I've talked to girls who have been exploited and simultaneously exposed to naked men in intimate areas of undressing. I've talked to girls who have been severely injured in their sports due to a male player on the opposing team."[22]

The Women's All-American Field Hockey Team, 1920s.

THE ETHICAL ISSUE

However, biological men identifying as women in order to compete against biological women is not just physically dangerous, it is also unethical, as Attorney General Bondi also noted:

"In February 2024, these are just a few examples, a biological boy started competing in women's ski races and cross-country races in Maine. He came in first in the 5k with a time that would have been 43rd among men. In February

2025, a biological boy won first place in the pole-vaulting competition in Maine's indoor track and field meets. He beat every other girl by a significant margin. That qualified him for regional championships that took a spot away from a young woman in women's sports. Shame on him."[23]

MALE HUNTERS AGAINST FEMALE GATHERERS

We have other sports examples of the unjustifiable disparity created when modern male "hunters" are put up against modern female "gatherers."

More for fun than serious competition, in 2017 the U.S. Women's National (Soccer) Team (USWNT) played an informal match against an all-male soccer team: the Football Club Dallas Academy Team. The all-women's team lost, with a final score of 5 to 2. Yet the all-male team was comprised entirely of teenage boys, all

Professional female archer Mrs. Philip Rounsevelle of Ashville, North Carolina, winner of the bronze medal, National Archery Tournament, 1920s.

under the age of 15.[24] In June 2023, in another unofficial game, an all-biological male British soccer team played an all-biological female U.S. soccer team. Again, the women's team lost. This time 12 to 0.[25]

Considering the physical superiority of men (hunters) compared to women (gatherers), should we be surprised at these one-sided outcomes?

OUR SEXUALLY DIMORPHIC BRAINS

Our discussion leads to two additional questions: If God/Nature designed male and female bodies for different functions, must not the brains of men and women also be designed for different functions? In other words, if there is such as thing as *physical* sexual dimorphism,[26] is there not also such a thing as *behavioral* sexual dimorphism? And if so, how does this relate to men playing in women's sports?

Sexual dimorphism is technically defined as the physical *and* behavioral differences between males and females (outside the reproductive organs) of the same species. This contradicts the Left's politically correct, socially engineered manifesto, which states that any cognitive and behavioral differences between men and women are the result of socialization and personal biases (that is, culture).

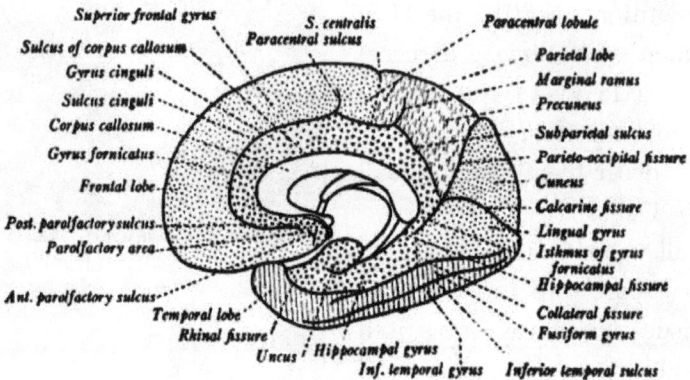

Diagram of the medial aspect of the human cerebral hemisphere. Male and female brains have striking differences in both physical appearance and behavioral programming. For example, while the two hemispheres of a woman's brain communicate information more rapidly—an adaptation to maternal and social concerns, the larger inferior-parietal lobule in a man's brain confers greater spatial, mathematical, and auditory abilities—an adaptation to hunting, defensive, and inventive concerns. Both modern male and female brains have their biopsychological origins in the hunting and gathering lifestyle of two of the earliest known humans: our distant ancestors *Homo habilis* and *Homo erectus*.

Impartial neuroscientific studies, however, show that there are biologically built-in distinctions between how male and female brains are designed, how they function, and how these factors influence behavior.[27] The *Journal of Neuroscience Research* goes as far as to say that the neural differences between men and women are not only permanent and lifelong, they exist "at all scales, from the genetic to the epigenetic, to the synaptic, cellular, and systems differences."[28]

Let us look at just a few of the more significant of these. Morphometric research shows, for instance, that:[29]

• On average, male brains are much larger and heavier than female brains, in most cases, at least 10 to 15 percent greater in size.[30]

• There is a higher percentage of white matter—which contains myelinated axonal fibers—in male brains than in female brains.[31] These fibers facilitate motor control and decision making, ideal for hunters and warriors.

• There is a higher percentage of gray matter in women than in men.[32] Gray matter, among other things, aids in the regulation of emotions, ideal for maternal and domestic tasks.

• Certain areas of the *corpus callosum* (the neural pathway that connects the two halves of the brain) are generally larger in women than in men. With more nerve fibers, this gives women a potential advantage in the speed at which they can transmit and receive (i.e., communicate) information,[33] ideal for maternalism: femininity, affection, nurturance, and motherliness.

• Men have thicker cortices in the right anterior temporal region (aids in the memory of visual and spatial information) and the orbitofrontal region (controls decision making)—which aids in hunting and defense. Women have thicker cortices in the right inferior parietal region (processes visual, tactile, and auditory information), the left ventral frontal region (controls social awareness), and the posterior temporal region (processes emotional, auditory, visual, memory, and language cues)—which aids in maternal and domestic concerns.[34]

• Maternalism: Women are more right-brained (controls emotions, gut feelings, intuition, communication, empathizing, and non-verbal thinking) than men; Paternalism: Men are more left-brained (controls analysis, systemizing, intellectual thought, abstract thought, and rational thought) than women.[35]

• The neurotransmitters serotonin and dopamine (both affect mood) are found in higher concentrations in females than in males—being happy, calm, and well-rested aids women during the rigors and pressures of pregnancy and child-rearing.[36]

Organized women's outdoor recreation club, 1920s.

We now come to the most important facts in this entire list, and the ones most closely associated with both established sexually dimorphic science (namely, that the two sexes have very different bodies as well as very different brains, hardwired skills, and behaviors), and one of the major themes of this book—namely, that the average man is innately and physically superior to the average female:

• *Men are better at processing abstract and spatial tasks; women are better at processing social and verbal tasks.*[37] In other words, men have a greater proclivity for manipulation of physical objects, intangible thought, and mechanical design than women, explaining why, *on average*, young boys prefer playing war games with other boys, while, *on average*, young girls prefer playing with dolls and other girls.

For skeptics, consider the research showing that this last statement is true despite how progressive or traditional a country is when it comes to defining sex roles—further validating men's universal obsession with tools and mechanical devices. And this turns out to be true not only among we human primates: When given a choice, male monkeys prefer playing with toy trucks over any other type of toy.[38]

Start of a women's cross-country ski race, 1927.

MANPOWER

It is clear that human males, with their much higher levels of testosterone (a sex hormone that has a masculinizing effect), massive powerful bodies, and spatially-oriented neural systems, have been hardwired principally for a big-game hunting, defensive, patriarchal (father-centered) mode of life.[39]

Women's biology, however, shows the opposite: With their much higher levels of estrogen (a sex hormone that has a feminizing effect), more gracile physical structure, and keen social and verbal skills, they are primarily hardwired for childbearing, (group) childcare, and gardening; that is, for a matriarchal (mother-centered) mode of life.[40]

Thus, while men and women are equals, we are not the same. For due to God/Nature, we are biologically and psychologically designed for different purposes, men as physically strong father-hunters, women as psychologically empathetic mother-gatherers—even though a majority of us are no longer living as hunter-gatherers. For it is a well-known scientific fact that sexual dimorphism does not always equate with how an animal currently lives, but more specifically how its ancestors lived.[41]

Let me propose the following analogy. A 21st-Century human is like a steam-powered engine installed in the body of a modern electric car: Outside we look, act, speak, and think "modern." But at heart we remain prehistoric creatures, evolutionarily hardwired for two very different biopsychological roles—and the science of neurology provides indisputable proof. As one neurobiologist has unequivocally stated:

> "The human brain is a sex-typed organ with distinct anatomical differences in neural structures and accompanying physiological differences in function."[42]

It is these very evolutionary biological differences in the brains—and by extension the physical bodies—of human males and human females that justify keeping biological men out of biological women's sports. In a recent interview former professional racing driver Danica Patrick put the matter this way:

"There's just some clear differences. So when it comes to strength and pure potential, there's no question that guys are able to achieve more than women."[43]

Sexual differentiation and dimorphism in the ancient world: Ancient Egyptian statue of King Menkaure and his queen, circa 2490 B.C. Notice the masculine placement of the king's hands and arms (by his side), and the feminine placement of the queen's hands and arms (around the king). Also take note of the hands themselves: The king's hands are closed, the queen's hands are open. As for the feet, the king's are positioned in an aggressive stance, the female's in a passive stance. Lastly, the king is shown, realistically, as physically larger, taller, stronger, and broader than the queen. All of these traits are archetypal symbols of the Masculine and Feminine Principles.

Miss Merion Hokins, National Golf Champion for 1921, executing the most difficult shot in polo: the off-side stroke.

CHAPTER TWO
SEXUAL DIMORPHISM
THROUGH TIME

D ESPITE THE CLAIMS OF THE misinformed to the contrary, it is clear from biology alone that the two sexes were designed by God/Nature for very different roles, and that the two brains at the center of these two different roles were specifically programmed for separate and precise tasks: the male for hunting and survival (and the many ancillary roles that are connected to these skills), the female for motherhood and gardening—and perhaps small-game hunting (and all of the ancillary roles that go with these skills). From out of these two evolutionary imperatives grew the hyper protective male brain and the hyper maternal female brain, brains that have not changed substantially in hundreds of thousands of years.

DECREED BY GOD/NATURE, DEFINED BY LOGIC, CONFIRMED BY SCIENCE
The acknowledgment of this sociobiological, sex-typed, two-gender blueprint is not new. It was well-known to the ancients and was set down, for example, in the Judeo-Christian Bible thousands of years ago:

> "So God created man in his own image, in the image of God created he him; male and female created he them."[44]

There is nothing here about hundreds or even dozens of sexes or genders. Only two.

Centuries later, Jesus reinforced God's male-female dictum during the following exchange with the Pharisees:

> And it came to pass, that when Jesus had finished these sayings, he departed from Galilee, and came into the coasts of Judaea beyond Jordan. And great multitudes followed him; and he healed them there. The Pharisees also came unto him, tempting him, and saying unto him: "Is it lawful for a man to put away his wife for every cause?" And he answered and said unto them, "Have ye not read, that he which made them at the beginning made them male and female, and said, 'For this cause shall a man leave father and mother, and shall cleave to his wife: and they twain shall be one flesh?' Wherefore they are no more twain, but one flesh. What therefore God hath joined together, let not man put asunder."[45]

By deemphasizing and even obliterating the lines demarcating the biologically predetermined sex assigned to human males and females at birth, modern progressives are not only going against God (i.e., both traditional Western religion and spirituality), they are also running counter to millions of years of biological evolution—as we shall see shortly.

HOW EARLY CULTURES ACKNOWLEDGED OUR SEX DIFFERENCES

Some 2,300 years ago, in an ancient parallel of the Torah, the Chinese developed the concept of yin-yang, today one of the foundational pillars of Chinese culture. What we have here in this ancient bisexual mandala is the idea of *two opposing energies that compliment one another*, like the two poles of an electric battery: one, the cathode pole, is

As the ancient Asian yin-yang symbol reveals, recognition of the Masculine Principle and the Feminine Principle—and how, though separate, they function as one unit—dates back thousands of years into the mists of prehistory.

positive (male), and marked with a plus sign; the other, the anode pole, is negative (female), and marked with a minus sign. The cathode and anode terminals are quite literally polar opposites. Yet "opposites attract," and, just as in a real battery, it is this relationship (attraction) that creates energy (life).

Even prior to the dawn of history humans recognized this natural division-compatibility archetype between the two sexes, even naming them the "Male Principle" and the "Female Principle." The former, the yang or male energy, has always typically been symbolized by one or more of the following terms and characteristics:

light
external
hot
dry
positive
daytime
hard
odd
logic
right
summery
upward
fast
assertive
energetic
rigid
large
spiritual
heaven

The great Greek Father-Sun-god Zeus (known in ancient Rome as Jupiter), Western personification of the Male Principle.

In contrast, the latter, the yin or female energy, has nearly always been symbolized by the following terms and characteristics:

dark
internal
cold
moist
negative
nighttime
soft
even
intuition
left
wintery
downward
slow
passive
flexible
small
material
earth

The great Greek Mother-Moon-goddess Hera (known in ancient Rome as Juno), Western personification of the Female Principle.

More commonly, yang (male) energy is best known for being symbolized as the sun, as paternal, as patriarchal, as fire, as wind, as spiritual rebirth, and as the heavenly father. However, Yin (female) energy has been most notably personified as the moon, as maternal, as matriarchal, as water, as rain, as physical death, and as the mother earth.

This eternal immutable male-female polarity has been codified, honored, and even worshiped, since prehistoric times, the result being a global pantheon of millions of gods and goddesses, which in turn has inspired two different but intricately interlaced religious fields: *theology*, the study of god (yang energy), and *thealogy*, the study of goddess (yin energy).[46]

We have on the yang side, for example, such father-gods and sun-gods as Zeus, Helios, Jupiter, Odin, Yahweh, Jehovah, Brahma, Sol, Ah Kin, Marduk, Baal, Horus, Ahura Mazda,

Phoebus, Belenos, Quetzalcoatl, Nanahuatzin, El, Surya, Christus, Ostara, Aten, Mithra, Hammon, Mahes, Lug, Apollo, Xu Kai, Marduk, Tawa, Isten, Ogma, Aton, Huiracocha, Malakbel, Simigi, Lugh, Hvar, Osiris, Tonatiuh, Bel, Cao Dai, Indra, Kinich Ahau, Izanagi, Papas, Hercules, Dielli, Inti, Nahhundi, Ra, Harachte, Mog Ruith, Istanu, Palk, Mihr, and Utu.

On the yin side we have such mother-goddesses and moon-goddesses as Juno, Brigid, Aditi, Mandulis, Luna, Demeter, Mahina, Frigg, Cybele, Artemis, Hlodyn, Selene, Hena, Tinnit, Asherah, Devana, Anumati, Hathor, Caelestis, Ilazki, Anahita, Myeongwol, Chia, Athena, Lakshmi, Izanami, Pasiphae, Ningal, Durga, Xochiquetzal, Mari, Mary, Isis, Heng E, Astarte, Kuu, Ammavaru, Cihuacoatl, Ishtar, Marama, Atargatis, Oya, Gauri, Astroarche, Ixchel, Mama, Gaia, Coyolxauhqui, Hina, Diana, Kali, Freya, Nammus, Arinna, Mama Killa, Dea Matrona, Ratih, and Bendis.[47]

Star of David, a mystical symbol of the union between spirituality (male) and materiality (female).

Every religion, in fact, has acknowledged and celebrated the male-female polarity, in many cases even establishing it as the very symbol of their faith. Out of ancient Judaism, for example, came the Star of David. Also known as the Magen David, it is comprised of two interlocking triangles: the downward pointing triangle represents the Masculine Principle, piercing Mother Earth below; the upward pointing triangle represents the Feminine Principle, receiving the Heavenly Father from above.[48]

Naturally, ancient Christianity also symbolically honored the universal male-female polarity as well: The Christian cross is made up of an "active" vertical beam, which points upward, symbolizing the heavenly Male Principle; a "passive" horizontal beam overlays it, representing the earthly Female Principle.[49]

In each case we have a representation of the male-female polarity archetype: one Asianized, one Judacized, and one Christianized, all three being cryptic emblems of the ancient rite known as hierogamy, or more popularly, the Hieros Gamos: the "Sacred Marriage" between male and female—which Plutarch revealingly called "the basis of all mysteries."[50]

On a deeper more esoteric level the Hieros Gamos represents the union of heaven and the earth, the union of the sun and the moon, the union of God and Man, the union

Christian Cross, an esoteric symbol of the sacred marriage between heaven (male) and earth (female).

of the soul and the body, and the incarnation of the spiritual (plane) on the material (plane). These numenistic "unions" are known variously throughout the world as spiritual enlightenment, cosmic consciousness, self-actualization, nirvana, self-realization, individuation, and samadhi.

BROTHER SUN, SISTER MOON

Christian mystic St. Francis of Assisi played upon these divine male-female themes in his classic 1225 poem, "The Canticle of Brother Sun and Sister Moon" (which was, in turn, artistically depicted in the 1972 Franco Zeffirelli film, *Brother Sun, Sister Moon*):

> Most high, omnipotent, good Lord,
> Praise, glory and honor and benediction all, are Thine.
>
> To Thee alone do they belong, most High,
> And there is no man fit to mention Thee.
>
> Praise be to Thee, my Lord, with all Thy creatures,
> Especially to my worshipful brother sun,
> He which lights up the day, and through him dost Thou
>   brightness give;

And beautiful is he and radiant with splendor great;
Of Thee, most High, signification gives.

Praised be my Lord, for sister moon and for the stars,
In heaven Thou hast formed them clear and precious and fair.

Praised be my Lord for brother wind
And for the air and clouds and fair and every kind of weather,
By the which Thou givest to Thy creatures nourishment.

Praised be my Lord for sister water,
The which is greatly helpful and humble and precious and pure.

Praised be my Lord for brother fire,
By the which Thou lightest up the dark.
And fair is he and gay [i.e., happy] and mighty and strong.

Praised be my Lord for our sister, mother earth,
The which sustains and keeps us
And brings forth diverse fruits with grass and flowers bright.

Praised be my Lord for those who for Thy love forgive
And weakness bear and tribulation.

Blessed are those who shall in peace endure,
For by Thee, most High, shall they be crowned.

Praised be my Lord for our sister, the bodily death,
From the which no living man can flee.
Woe to them who die in mortal sin;

Blessed those who shall find themselves in Thy most holy will,
For the second death shall do them no ill.

Praise ye and bless ye my Lord, and give Him thanks,
And be subject unto Him with great humility.[51]

Reading between the lines we discern a millennia old, globally accepted tradition: The male-yang pole and the female-yin pole are fixed, changeless, and inexorable, bound together as separate but unified energies, biopsychologically preprogrammed to work harmoniously but individually at their own separate tasks and symbiotic roles. Whether an average male or an average female, whether a prehistoric human or a modern human, we are what we are: human primates biologically destined to play the differentiated yet integrated roles that we have inherited from our days as Stone Age hunter-gatherers.

BIOLOGICAL REALITY VS. ABSTRACT THEORY

Does any of this mean that men cannot be maternal and women cannot be paternal? Of course not, and, as is obvious from history, numerous examples could be cited that demonstrate this.

It is merely to say that, because the average man is physically stronger than the average woman, he is more likely to outperform the average woman when it comes to physical sports. On the other hand, the average woman is more maternal than the average man, making it more likely that she will outperform the average man when it comes to childcare and domestic concerns. Before feminism infiltrated and appropriated the legal system, this was, after all, the original reason that divorce courts traditionally awarded custody of children and the house to single mothers.[52]

An early magician's drum from Lapland with sun and moon symbols on it, evidence that Scandinavians have been worshiping the Male and Female Principles for over 1,000 years.

Being socially adept creatures who are the products of both Nature and Culture, to some extent we certainly have the power to override our evolutionary programming: Though it is hardwired

Female skaters, practicing for the U.S. Figure Skating Championship, 1920s.

into their nature, men do not have to hunt or protect; and though it is hardwired into their nature, women do not have to be maternal or bear children. Conversely, men can be maternal and care for children, and women can be protective and hunt and fish.[53]

However, the intellectual choices we make do not alter the evolutionarily hardwired proclivities embedded in us by God/Nature. These are permanent aspects of who we are, as is apparent when we stand the average powerfully built male next to the average more delicately built female. After carefully studying our countless built-in sex-typed differences (corroborated by both science and religion), only a simpleton would claim that the two sexes are perfectly equal in every respect, or that one is overall superior to the other.

The magnetically charged male-female polarity cannot be erased. It is a permanent biopsychological fixture in our species. Indeed, our very sense of the biological sex differences in men and women seems to be built-in, for recognition of the male-female polarity is universal, with studies showing that in the majority of cultures both men and women prefer masculine males and feminine females[54]—meaning feminine males and masculine females are unpopular worldwide, a preference that goes beyond race, religion, nationality, and ethnicity.

The all-powerful attraction between masculine energy and feminine energy is omnipresent, which is why it is even found in homosexual relationships: Among lesbian couples (biological women) a feminine female is called a "femme," her masculine partner is called a "butch"; among gay couples (biological men) a feminine male is called a "femboy," his masculine partner is called a "masc." Whatever our chromosomal arrangement, or our sexual proclivity, identity, or preference then, the male-female polarity holds true, cutting across all sexual demarcations, artificial barriers, social lines, and cultural boundaries.

Some will assert that these views are hypothetical; and it is true that one might base them solely on common sense, knowledge, and education. However, they are far from being simply theories. Besides the many ancient deities, myths, religious doctrines, sacred scripture, statuary, and art that emphasize male-female sex differences, in more modern times there are centuries of solid scientific research, study, and observation, by some of the world's most brilliant minds, to support them. This stance has even been organized into a relatively new branch of science, sociobiology, which is defined as "the scientific study of the biological, especially ecological and evolutionary, aspects of social behavior in animals and humans."[55] I am speaking here, not of abstract theory, but of biological reality.

SEXUAL DIMORPHISM IN NONHUMAN ANIMALS

In short, unlike the philosophy of radical feminism, which teaches that "women are superior to men,"[56] according to the sociobiological model there can be no absolute superiority or inferiority when its comes to the two sexes. Here, men and women are truly equals—though, depending on our sex, we each have unique talents and skills assigned to us by God/Nature at conception. Biology (though not necessarily psychology) is always predetermined by mathematically precise immutable laws; and this applies to both human animals and nonhuman animals.

For instance, sexual dimorphism, the hunter-gatherer imperative that sculpted men into stronger, bigger, and faster beings than women, far from being merely a phenomenon found in *Homo sapiens*, is rife across the animal kingdom, with some of the more common and conspicuous examples being:

- Chimpanzees.
- Gorillas.
- Bonobos.
- Mandrills.
- Hamadryas baboons.
- Proboscis monkeys.
- Fan-throated lizards.
- Chickens.
- Mountain lions.
- Horses.
- Elk.
- Orcas.
- Sea otters.
- Wild turkeys.
- Olympic marmots.
- Capuchin monkeys.
- Grizzly bears.
- Black bears.
- Polar bears.
- Bighorn sheep.
- Wolves.
- Rhinoceroses.
- Sperm whales.
- Pheasants.
- Foxes.
- Rhesus macaques.
- Green anoles.
- House sparrows.
- Belugas.
- Elephant seals.
- Giraffes.
- Badgers.
- Mice.
- Orange tip butterflies.
- Bottlenose dolphins.
- Gharials.
- Hippopotamus.
- Pilot whales.
- Water buffalo.
- Blue tits.
- Spinner dolphins.
- African lions.
- Nilgai.
- Asian paradise flycatchers.
- Mandarin ducks.
- Zebras.
- Rats.
- Guppies.
- Wolverines.
- King cobras.
- Narwhals.
- Orangutans.
- Great Indian bustard.
- Peafowls.
- Wolves.
- Bison.
- Pronghorn antelope.
- Sage grouse.
- Leopard seals.
- Wildebeest.
- Rattlesnakes.
- Monitor lizards.
- Warthogs.
- Lesser floricans.

Sexual dimorphism is such an inherent evolutionary aspect of the animal kingdom that it is found, not only in *all* humans, from prehistoric to modern,[57] but in most animal species (at least 70 percent),[58] including fish and insects. It can even be traced as far

A young male gorilla, a fellow primate with one of the highest percentages of male-biased sexual size dimorphism. A close cousin of ours, we share not only 98 percent DNA with one another, but numerous impressive differences between males and females as well.

back as the archosaurs (dinosaurs and phytosaurs),[59] dating to hundreds of millions of years ago.[60]

As for mammals specifically, scientists have found that they are generally dimorphic, that on average males are larger than females, that male-biased size dimorphism exists in a majority of species, and that in over 45 percent of mammal species males are at least 10 percent larger or more.[61]

Clearly sexual dimorphism is real and has definitive functions and purposes.

The intelligent, objective, and inquisitive want to know why, and will impartially search for, and accept, the answer—whatever it turns out to be. This is the scientific method, pure and simple. The emotional, subjective, and apathetic, however, will ignore the facts, and impose their own feelings-based ideologies on biology. This, in stark contrast to the scientific approach, is the ideological method, one that goes against both God (spiritual proof) and Nature (physical proof).

MORE EVIDENCE: A STUDY OF HUMANS OVER TIME

Lastly, if more proof is required to support our view that biological men should be barred from biological women's sports, we need only look at human morphology. During the Bronze Age (3300 B.C. to 1200 B.C.), for example, men were as much as 8 percent taller than women, about the same as in modern humans,[62] proof that it is not just modern humans who are sexually dimorphic. Human males were already larger and stronger, that is, more powerful and more potentially dangerous, than human females many thousands of years ago.

Known by anthropologists as "The Stag Hunt," this 7,000 year old mural painting in the "Cueva de los Caballos" near Albocacer, Castellon, Spain, clearly demonstrates the hunting lifestyle of early Europeans. After the hunt the men skinned and cut up their quarry, then transported it back to camp where the group's females took over, preparing, storing, and cooking the meat, bones, and entrails. These carnivore-heavy meals were often mixed with small game and vegetable matter collected by the females. The bodies and minds of modern men and women retain the evolutionary imprint of millions of years of the hunting-gathering lifestyle, making the participation of biological men in biological women's sports one-sided, inequitable, discriminatory, and scientifically untenable.

Yet, as we are about to see, this general metric holds true for *all* humans, not just modern ones, going back millions of years. In fact, sexual dimorphism between male and female hominins seems to increase as we go backward in time,[63] with one ancestral relative, the male Australopithecine, being as much as 50 percent larger than his female counterpart.[64] Let us make note of the fact that the genus *Australopithecus* lived between 4 million and 2 million years ago.

With these enlightening stats in mind, let us look at ourselves more closely for a moment, including the average height and weight, as well as the average brain size, of today's American men and women:

• Modern American males have an average height of 5 feet 9 inches, an average weight of 199.8 lbs, and an average brain size of 1,260 cubic centimeters (cc).
• Modern American females have an average height of 5 feet 4 inches, an average weight of 170.8 lbs, and an average brain size of 1,130 cc.

From this we see that American men, on average, are consistently 5 inches taller and 29 pounds heavier than American women (that is, overall about 9 percent bigger),[65] with brains that are 130 cc larger.

We will note that although the average weight for both sexes has gone up significantly over the past 65 years, this is a superficial social effect, in nearly all cases a result of poor eating choices and lack of exercise. Either way, males still retain their more robust bodies, females still retain their more gracile bodies—an evolutionary adaptation to our two very different but compatible roles as part of the hunter-gatherer lifestyle.

Nothing has changed these differences, even after hundreds of thousands of years; and nothing is going to change them at any time in the near future. This is easily proven by simply looking back in time, at both the earliest and most recent forms of *Homo*: our human family.

THE NEANDERTHAL PEOPLE

Let us start with, not an ancestor, but a cousin of ours: Our closest extinct relative, *Homo neanderthalensis* ("Man from the Neander Valley").[66] This archaic human species, with whom we share DNA (the result, not of evolution, but of interbreeding), lived from 400,000 to 40,000 years ago:

A 1905 (outdated) illustration of two Neanderthal men.

- Neanderthal males had an average height of 5 feet 5 inches, an average weight of 143 pounds, and an average brain size of 1,600 cc.
- Neanderthal females had an average height of 5 feet 1 inches, an average weight of 119 pounds, and an average brain size of 1,300 cc.

Thus, Neanderthal men, on average, were 4 inches taller and 24 pounds heavier than Neanderthal women, with brains that were 300 cc larger. Though we did not evolve from Neanderthals, they share the same sexual dimorphism that is common to all known hominins, both recent and archaic.

THE CRO-MAGNON PEOPLE

Next in line, we must look at the first modern human, Cro-Magnon, a prehistoric version of we ourselves: *Homo sapiens sapiens* (in other words, we are modern day Cro-Magnons). This human group—named for the French cave in which they were first discovered in 1868—lived from about 45,000 years ago to 10,000 years ago:

- Cro-Magnon males had an average height of 5 feet 9 inches, an average weight of 143 pounds, and an average brain capacity of 1,600 cc.
- Cro-Magnon females had an average height of 5 feet 4 inches, an average weight of 119 pounds, and an average brain capacity of 1,514 cc.

Two Cro-Magnon men (modern humans) painting a symbolic representation of a recent successful big game hunt on a cave wall.

Thus, Cro-Magnon men, on average, were 5 inches taller and 24 pounds heavier than Cro-Magnon women, with brains that were 86 cc larger.[67]

HEIDELBERG MAN

Reconstruction of Heidelberg Man, a prehistoric hunter-gatherer from whom we inherited much of our modern sexual dimorphism.

For more convincing metrics let us look at who is currently believed to be our immediate evolutionary ancestor, *Homo heidelbergensis* ("Heidelberg man"). Also the probable common ancestor of our cousins the Neanderthals, this archaic human species thrived between 600,000 and 300,000 years ago:

- Heidelberg males had an average height of 5 feet 9 inches, an average weight of 136 pounds, and an average brain size of 1,270 cc.
- Heidelberg females had an average height of 5 feet 2 inches and an average weight of 112 pounds, and an average brain size of 1,130 cc.

Paleolithic hunt: Stone Age men stalking a wooly rhinoceros. Thanks to our hunting and gathering ancestors, modern men are still perfectly physically and psychologically adapted to this mode of life—even though it is no longer necessary for a majority of men.

Thus, Heidelberg men, on average, were 7 inches taller and 24 pounds heavier than Heidelberg women, with brains that were 140 cc larger.

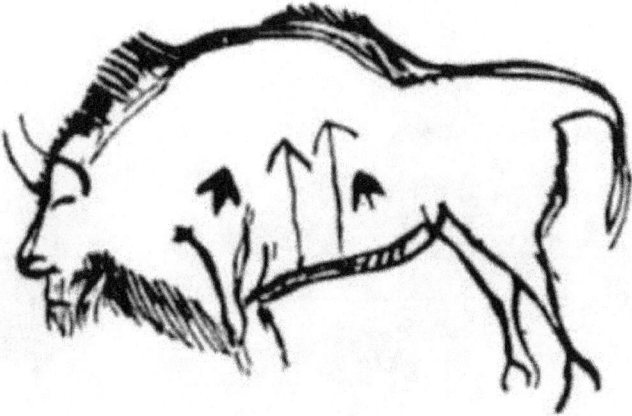

Cave art, from the cavern of Niaux, France, showing spears and arrowheads embedded in the side of a steppe bison—14,000 year old proof of prehistoric human hunting, and by extension, food-gathering.

UPRIGHT MAN

Next we will examine the immediate ancestor of Heidelberg Man and the secondary ancestor of modern humans: *Homo erectus* ("upright [walking] man"). This primitive hominid lived from 1.6 million years ago to around 250,000 years ago (note: currently there is no reliable data on the cranial capacities of either male or female *Homo erectus*):[68]

- *Homo erectus* males had an average height of 5 feet 11 inches, and an average weight of 145 pounds.
- *Homo erectus* females had an average height of 5 feet 3 inches and an average weight of 123 pounds.

Thus, *Homo erectus* men, on average, were 8 inches taller and 22 pounds heavier than *Homo erectus* women—or 13 percent bigger.[69]

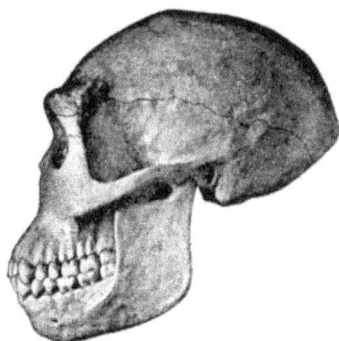

H. *erectus* skull, side view.

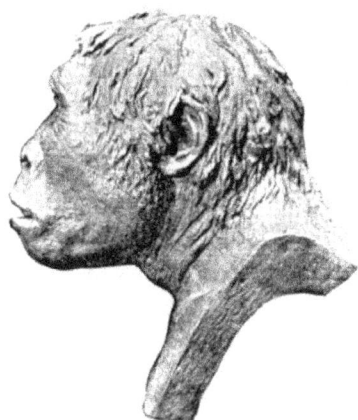

H. *erectus* head, side view.

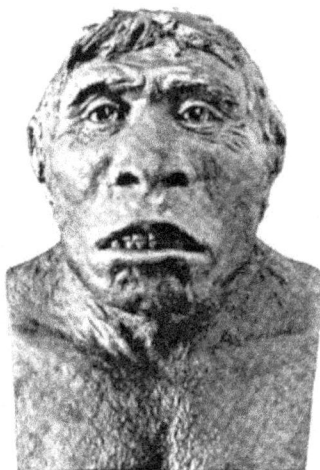

H. *erectus* head, front view.

H. *erectus* bust.

Four views of *Homo erectus*, formally known as Java Man or *Pithecanthropus erectus*—another early ancestral hunter-gatherer.

HANDY MAN

Let us touch on another human relative, the immediate ancestor of *Homo erectus*, the secondary ancestor of Heidelberg Man, and the tertiary ancestor of modern humans: *Homo habilis* ("handy man"). The earliest known member of the genus *Homo*, this extinct human species lived from 2.4 to 1.5 million years ago (note: currently there is no reliable data on the cranial capacities of male and female *Homo habilis*):

- *Homo habilis* males had an average height of 5 feet 2 inches, and an average weight of 114 pounds.
- *Homo habilis* females had an average height of 4 feet 1 inches and an average weight of 70 pounds.

Thus, *Homo habilis* men, on average, were 1 foot 1 inch taller and 44 pounds heavier than *Homo habilis* women—or 26 percent bigger.[70]

THE SOUTHERN APE

Let us examine one final species: *Australopithecus afarensis* ("southern ape from Afar [Ethiopia]"). Even greater dimorphism has been noted in this relative, and possible prehistoric ancestor, of *Homo*,[71] one who lived between 3.85 and 2.95 million years ago.[72] A study of *A. afarensis* fossil footprints, for instance, has revealed a probable "strong" sexual dimorphism, one equal to that of modern gorillas, with males being twice the size of females.[73] Another study estimates a male *A. afarensis* weight of 149 pounds, and a female *A. afarensis* weight of 79 pounds,[74] showing the greatest sexual size dimorphism between male and female of any hominin we have looked at. If these educated guesses turn out to be true, *A. afarensis* men, on average, were 70 pounds heavier than *A. afarensis* women, with the male being as much as 44 to 50 percent bigger than the female.[75]

Prehistoric all-male hunting band closing in on a wooly mammoth.

GOD & THE GAME OF CHANCE

Reviewing these indisputable facts, let us agree with Einstein, who, in 1926, said: "God does not play dice." In other words, since God is the architect of Nature, nothing in Nature can be accidental. All design is intentional, functional, and preprogrammed, a truth that can be observed from the smallest quark to the largest supergalaxy: Modern men began in prehistoric times as fearless hunters and powerful protectors; modern women began in prehistoric times as resourceful gatherers and maternal childcare providers. In spite of our 21st-Century guise, nothing has altered these two biological truths in the modern era. This could not be made any clearer than from, as we have seen, a study of the sexual dimorphism of our own kind—both past and present.

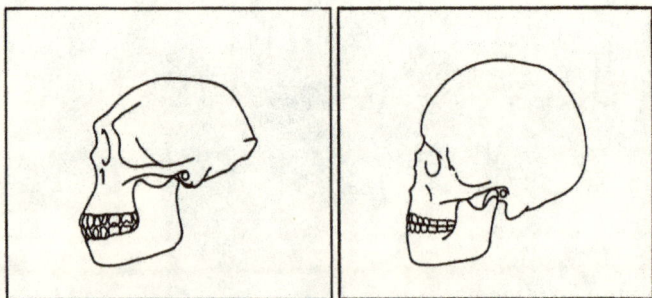

Comparison of a *Homo erectus* skull (left) and a modern human skull (right).

We must also consider the fact that primate species which, unlike humans, are not hunter-gatherers, but only gatherers, are not generally sexually dimorphic. When it comes to diet, the gibbon, for example, is primarily a food-gathering primate that only occasionally hunts living prey (like small birds and insects). Unsurprisingly, it is monomorphic, meaning that the two sexes share almost identical physical characteristics. Why?

Since it relies on only one primary feeding habit, namely gathering, there is little or no competition for food. Therefore, there is no need for the type of division of labor found in sexually dimorphic animals, like humans, where one sex (the robust male) has mentally and physically evolved to hunt and kill big game, while the other sex (the gracile female) has evolved mentally and physically to care for children and engage in gardening, and perhaps occasionally small game hunting.[76]

What does this mean for biological men claiming to be females so they can participate in biological women's sports? Simply put, it means that biological female athletes will always be unfairly physically outmatched, an overt violation of both women's civil rights and sports ethics, and a clear and present danger to women's physical safety and psychological health.

Reconstruction of Cro-Magnon Man: *Homo sapiens*, the earliest modern human, with whom we share identical sexual dimorphism.

Miss Beatrice Lougheran, national figure skating champion, just one of thousands of professional women skaters who competed in the early 20th Century.

CHAPTER THREE
BIOLOGICAL SEX
VS. GENDER IDENTITY:
SOLUTIONS

BASED ON A MULTITUDE OF scientific disciplines, I believe I have made my case as to why biologically natal males should not be allowed to compete in women's sports, and by extension be barred from female-only spaces. Women's sex-based rights must not only be respected. They must be permanently codified into law for the protection and safety of females of all ages from this time forward, covering law, language, and culture.[77] Adding the 1972 28th (or Equal Rights) Amendment to the U.S. Constitution would be extremely helpful in this regard. But this has yet to occur, despite numerous individual states ratifying it over the past five decades.

The importance of establishing sex-segregated sports, that is, using biological sex rather than gender identity, as the basis for protecting women's sports is driven home by the fact that trans ideologies, programs, and goals are anti-female, misogynistic, and retrogressive, further eroding women's already fragile civil rights when it comes to sports.

Moreover, trans policies only serve to embroil biological women in an issue that most have no interest in, and one therefore that they should not have to deal with. Trans people (and their supporters) are responsible for the problems that they themselves have created. These should not be imposed on others, in particular biological women. If you violate someone's constitutional, civil, or religious rights, not to mention inconvenience, belittle, mistreat, threaten, abuse, and even physically injure them, why would you expect them to assist you? In other words, trans activists expect the very individuals they are hurting (biological women and their families) to support them.

Center: Mrs. J. W. Connors winning the Secretary of the Navy Cup Race in her boat *Miss Okeechobee*, 1920s.

Women's right activist and former U.S. pro swimmer Riley Gaines has stated that trans policies are actually a "betrayal" of women's rights:

> "Allowing males to compete on a women's team is and always will be unfair, and the burden should not be placed on female athletes to convince their schools to accept this [as] scientific reality. But that's where it is being placed. And we as women, we're being silenced. Our universities and institutions are gaslighting and emotionally blackmailing us to make us the likely oppressors."[78]

Yet when biological female athletes speak out against allowing biological men to play in women's sports they are often told by educational administrators to "seek counseling," and that they need to learn to "accept" pro-trans ideology.[79]

Of course, transgenderism would not exist at all if its

proponents accepted and followed the laws of science, and more specifically those related to cytogenetics. Instead, they have chosen the self-deceptive route, ignoring a biological reality that even children understand: One's natal sex cannot be changed. The arrangement of our sex chromosomes is permanent. We are born either male or female (see endnote 80 on intersex individuals), and no operation, chemical, law, or ideology will ever alter this fact. In fact, in all of world history no one has ever changed their sex. And no one ever will. It is impossible.

Scottish lacrosse team, 1920s. The first international women's matches began in 1912.

At the other end of the spectrum is transgenderism: a fluid, ever-changing, idiosyncratic, highly personal, wholly subjective and individualistic ideology that is in turn based on feelings and only feelings—that is, how and what one *feels* about his or her gender, irregardless of what chromosomal sex pattern (XX or XY) they were born with. This is why there is no medical test for transgenderism: Whether one is transgender or not transgender is based solely on one's personal impressions, thoughts, opinions, sentiment, views, and emotions. There is no science behind it whatsoever.

While sex is a biological issue, gender is a psychological one, for, to repeat, gender identity is based on one's interior sense of what he or she *feels* they are or should be *at that moment*: male, female, both, or neither. As such, gender self-identifying can change from week to week, day to day, even moment to moment. Is it any wonder that what is now known as gender dysphoria was originally called "gender identity disorder," with its own listing in the *Diagnostic and Statistical Manual of Mental Disorders*? Why was the name changed? Was it due to advancements in psychiatry or the influence of progressive politics, or both?

Women onlookers watching a female rowing team practicing their sport at Cornell University, 1927.

Yet one thing remains the same, before, now, and forever after: A man is born, not made; a woman is born, not made. As I have said, our sex is decreed by God/Nature, defined by logic, confirmed by science. Only gender can be fabricated on a whim—and it is being used as a tool to deconstruct Western society and as a weapon to hurt biological women. The pure subjectivity of transgenderism, seemingly purposefully introduced and provoked by some groups, is just one reason that there is so much misunderstanding and animosity surrounding this issue.[80]

In any case, throwing out the original proper definition of sex and replacing it with a false manmade one while completely ignoring scientific reality, solves nothing. This only exacerbates the many difficulties associated with pushing transgenderism on an unwilling and disinterested majority.

Girl sailor, Marblehead, Massachusetts, 1927.

PROTECTING WOMEN IN WOMEN'S SPORTS

As a women's sports advocate, what then is my solution?

The ultimate goal, as Attorney General Bondi has declared, is to "protect women in women's sports." More broadly speaking, all of women's sex-based rights must be protected.

In my opinion, to accomplish this we must begin by ending Title IX infractions while putting a stop to the nefarious erosion of women's civil rights. At the same time, transwomen (men claiming to "feel" like women) should be required to create their own sports leagues, teams, and events and be allowed to compete only against other transwomen.

Additionally, although the countless opportunities and benefits lost by biological women to transwomen over the years can never be truly recovered, the titles and awards won by transwomen athletes can. These, of course, should be returned to their rightful biological female winners, the names of transwomen "winners" should be struck from the record books, and these individuals themselves irreversibly disqualified. Furthermore, the U.S. government must consider stopping (even retroactively) federal funding from those institutions not complying with the law.

All-women's archery competition: Second Annual Tournament of the Southern California Archery Association, 1927.

All-girls and women's class in horsemanship at Camp Tegawitha, Mt. Pocono, Pennsylvania, 1927.

Accompanying these changes transwomen must be legally and permanently banned from biological women's intimate spaces, all which should be formally labeled "female-only facilities"—off limits to everyone except biological females. America must return to the long held, science-based tradition of single sex spaces. Transpeople must create their own private trans spaces, separate from sex-based spaces. To ensure both Federal civil rights law and Title IX compliance, the regulations surrounding these issues should be strengthened and enforced, while sports leagues and educational institutions must be brought into alignment with the law through pertinent legal measures.

Due to their self-created self-identification policies, it is looking more and more like transpeople will need to form their own separate sub-society within the larger societal framework. For the vast majority of conventional everyday people, that is, non-transpeople, have spoken loudly and clearly: They have no desire to modify traditional Western culture to fit a nontraditional communist ideology, one that not only displeases and inconveniences the majority of people, but which actually hurts them on numerous levels, from psychological and financial to educational and occupational.

To summarize, biological natal males have no place in and no right to participate in women's sports. By doing so they are taking advantage of what I have termed the "Hunter-Gatherer Principle," in turn violating all that it means to be a biological woman, as well as the inherent rights that go along with being born a biological woman.

In the end there is only one remedy for dealing with this delusionary ideology. As U.S. Attorney General Bondi has declared:

> "It's pretty simple: Girls play in girls' sports, boys play in boys' sports; men play in men's sports, women play in women's sport."[81]

These views, ideas, and policies are the correct, fair, legal, simplest, and most logical and honorable way to solve the many problems, troubles, and issues created by the rise of the communist-based ideology known as transgenderism.

Belgian Women's Fencing Team, 1927.

Women skiers at Mt. Holyoke College, South Hadley, Massachusetts, 1927.

IN SUMMATION

Despite perceptions to the contrary, there is good news for transactivists in all of this: In retaining traditional democratic Western values, laws, and views they will not be as negatively impacted as they would have were they allowed to overturn the traditional gender binary. In fact, protections against discrimination will continue, just as before. The right to identify as whatever gender you choose, whenever and however you choose, will not be and cannot be infringed. In fact, transgenderism may and will legally continue to operate across the U.S., fully protected by the 14th Amendment. The only caveat is that it must operate in its own sphere, and without violating the constitutional rights of non-transpeople—which is *99.4 percent of the American population.*[82]

Yes, we must learn to live alongside those we disagree with, alongside people even completely different than ourselves. When it comes to sex-based sports, a rational understanding and application of the Hunter-Gatherer Principle can and will help the world do just that.

The End

One of the world's premiere hunting and gathering peoples: the indigenous Shoshone of North America.

APPENDICES

APPENDICES

APPENDIX A

FEDERAL LAWS RELATED TO
SEX-BASED WOMEN'S SPORTS

TITLE IX: "No person in the United States shall, on the basis of sex, be excluded from participation in, be denied the benefits of, or be subjected to discrimination under any education program or activity receiving federal financial assistance." — EDUCATION AMENDMENTS OF 1972

5TH AMENDMENT: "No person shall be held to answer for a capital, or otherwise infamous crime, unless on a presentment or indictment of a Grand Jury, except in cases arising in the land or naval forces, or in the Militia, when in actual service in time of War or public danger; nor shall any person be subject for the same offence to be twice put in jeopardy of life or limb; nor shall be compelled in any criminal case to be a witness against himself, nor be deprived of life, liberty, or property, without due process of law; nor shall private property be taken for public use, without just compensation." — U.S. CONSTITUTION

28TH AMENDMENT: "Equality of rights under the law shall not be denied or abridged by the United States or by any State on account of sex." — THE "EQUAL RIGHTS AMENDMENT," PROPOSED TO THE U.S. CONSTITUTION

APPENDIX B

KEEPING MEN OUT OF WOMEN'S SPORTS
Executive Order of the President of the United States of America
47th U.S. Chief Executive, Donald J. Trump

B Y THE AUTHORITY VESTED IN me as President by the Constitution and the laws of the United States of America, and to protect opportunities for women and girls to compete in safe and fair sports, it is hereby ordered:

Section 1. Policy and Purpose. In recent years, many educational institutions and athletic associations have allowed men to compete in women's sports. This is demeaning, unfair, and dangerous to women and girls, and denies women and girls the equal opportunity to participate and excel in competitive sports.

Moreover, under Title IX of the Education Amendments Act of 1972 (Title IX), educational institutions receiving Federal funds cannot deny women an equal opportunity to participate in sports. As some Federal courts have recognized, "ignoring fundamental biological truths between the two sexes deprives women and girls of meaningful access to educational facilities." Tennessee v. Cardona, 24-cv-00072 at 73 (E.D. Ky. 2024). See also Kansas v. U.S. Dept. of Education, 24-cv-04041 at 23 (D. Kan. 2024) (highlighting "Congress' goals of protecting biological women in education").

Therefore, it is the policy of the United States to rescind all funds from educational programs that deprive women and girls of fair athletic opportunities, which results in the endangerment, humiliation, and silencing of women and girls and deprives them of privacy. It shall also be the policy of the United States to oppose male competitive participation in women's sports more broadly, as a matter of safety, fairness, dignity, and truth.

Sec. 2. Definitions. The definitions in Executive Order 14168 of January 20, 2025 (Defending Women from Gender Ideology Extremism and Restoring Biological Truth to the Federal Government), shall apply to this order.

Sec. 3. Preserving Women's Sports in Education. (a) In furtherance of the purposes of Title IX, the Secretary of Education shall promptly:

(i) in coordination with the Attorney General, continue to comply with the vacatur of the rule entitled "Nondiscrimination on the Basis of Sex in Education Programs or Activities Receiving Federal Financial Assistance" of April 29, 2024, 89 FR 33474, see Tennessee v. Cardona, 24-cv-00072 at 13-15 (E.D. Ky. 2025), and take other appropriate action to ensure this regulation does not have effect;

(ii) take all appropriate action to affirmatively protect all-female athletic opportunities and all-female locker rooms and thereby provide the equal opportunity guaranteed by Title IX of the Education Amendments Act of 1972, including enforcement actions described in subsection (iii); to bring regulations and policy guidance into line with the Congress' existing demand for "equal athletic opportunity for members of both sexes" by clearly specifying and clarifying that women's sports are reserved for women; and the resolution of pending litigation consistent with this policy; and

(iii) prioritize Title IX enforcement actions against educational institutions (including athletic associations composed of or governed by such institutions) that deny female students an equal opportunity to participate in sports and athletic events by requiring them, in the women's category, to compete with or against or to appear unclothed before males.

(b) All executive departments and agencies (agencies) shall review grants to educational programs and, where appropriate, rescind funding to programs that fail to comply with the policy established in this order.

(c) The Department of Justice shall provide all necessary resources, in accordance with law, to relevant agencies to ensure expeditious enforcement of the policy established in this order.

Sec. 4. Preserving Fairness and Safety in Women's Sports. Many sport-specific governing bodies have no official position or requirements regarding trans-identifying athletes. Others allow men to compete in women's categories if these men reduce the testosterone in their bodies below certain levels or provide documentation of "sincerely held" gender identity. These policies are unfair to female athletes and do not protect female safety. To address these concerns, it is hereby ordered:

(a) The Assistant to the President for Domestic Policy shall, within 60 days of the date of this order:

(i) convene representatives of major athletic organizations and governing bodies, and female athletes harmed by such policies, to promote policies that are fair and safe, in the best interests of female athletes, and consistent with the requirements of Title IX, as applicable; and

(ii) convene State Attorneys General to identify best practices in defining and enforcing equal opportunities for women to participate in sports and educate them about stories of women and girls who have been harmed by male participation in women's sports.

(b) The Secretary of State, including through the Bureau of Educational and Cultural Affairs' Sports Diplomacy Division and the Representative of the United States of America to the United Nations, shall:

(i) rescind support for and participation in people-to-people sports exchanges or other sports programs within which the relevant female sports category is based on identity and not sex; and

(ii) promote, including at the United Nations, international rules and norms governing sports competition to protect a sex-based female sports category, and, at the discretion of the Secretary of State, convene international athletic organizations and governing bodies, and female athletes harmed by policies that allow male participation in women's sports, to promote sporting policies that are fair, safe, and in furtherance of the best interests of female athletes.

(c) The Secretary of State and the Secretary of Homeland Security shall review and adjust, as needed, policies permitting admission to the United States of males seeking to participate in women's sports, and shall issue guidance with an objective of preventing such entry to the extent permitted by law, including pursuant to section 212(a)(6)(C)(i) of the Immigration and Nationality Act (8 U.S.C. 1182(a)(6)(C)(i)).

(d) The Secretary of State shall use all appropriate and available measures to see that the International Olympic Committee amends the standards governing Olympic sporting events to promote fairness, safety, and the best interests of female athletes by ensuring that eligibility for participation in women's sporting events is determined according to sex and not gender identity or testosterone reduction.

Sec. 5. General Provisions. (a) Nothing in this order shall be construed to impair or otherwise affect:

(i)   the authority granted by law to an executive department or agency, or the head thereof; or

(ii) the functions of the Director of the Office of Management and Budget relating to budgetary, administrative, or legislative proposals.

(b) This order shall be implemented consistent with applicable law and subject to the availability of appropriations.

(c) This order is not intended to, and does not, create any right or benefit, substantive or procedural, enforceable at law or in equity by any party against the United States, its departments, agencies, or entities, its officers, employees, or agents, or any other person.

(d)   If any provision of this order, or the application of any provision to any person or circumstance, is held to be invalid, the remainder of this order and the application of its provisions to any other persons or circumstances shall not be affected thereby.

U.S. PRESIDENT DONALD J. TRUMP
THE WHITE HOUSE
February 5, 2025

APPENDIX C

THE 1997 LANDMARK REPORT ON *Physical Activity and Sport in the Lives of Girls* from the President's Council on Physical Fitness and Sports tells us some of the steps we can take to enable girls to reach their full potential.

It tells us that we need to provide more quality school-based physical education for girls. It tells us that we need to encourage girls to get involved in sport and physical activity at an early age. It tells us that we need to challenge stereotypes that impede girls' participation in sports. And it tells us that we have made progress in some areas.

For example, the Title IX legislation enacted in 1972 has opened the doors for millions of girls to participate in school sports. Americans took enormous pride in the accomplishments of the 1996 Olympic gold medal female athletes in soccer, softball, swimming, track and field, gymnastics, basketball, and other sports. We need to build on that spirit and develop a national commitment to ensure that every girl receives the encouragement, training, and support she needs to develop and maintain an active lifestyle.

Increasing physical activity among girls is a formidable public health challenge, but the potential rewards are great: a more vigorous nation, better health and greater leadership opportunities for girls, prevention of premature death and unnecessary illness, and a higher quality of life for our citizens.

I strongly encourage all Americans to join us in this effort.

Donna E. Shalala
Secretary of Health and Human Services
1997

APPENDIX D

UNITED KINGDOM SUPREME COURT JUDGMENT ON THE LEGAL DEFINITION OF A WOMAN

JUDGMENT

For Women Scotland Ltd (Appellant) *v* The Scottish Ministers (Respondent)
before
Lord Reed, President
Lord Hodge, Deputy President
Lord Lloyd-Jones
Lady Rose Lady Simler.

JUDGMENT GIVEN ON 16 April 2025

Heard on 26 and 27 November 2024

Appellant Aidan O'Neill
KC Spencer Keen
(Instructed by Balfour + Manson LLP (Edinburgh))

Respondent Ruth Crawford KC
Lesley Irvine
(Instructed by Scottish Government Legal Directorate)

Intervener – Sex Matters
Ben Cooper KC
David Welsh
(Instructed by Gilson Gray LLP (Edinburgh))

Intervener – Scottish Lesbians; The Lesbian Project; LGB Alliance (written submissions only)
Karon Monaghan KC
Beth Grossman
(Instructed by Doyle Clayton (London))

Intervener – (Equality and Human Rights Commission)
Jason Coppel KC
Zoe Gannon
(Instructed by Equality and Human Rights Commission)

Intervener – Amnesty International UK (written submissions only)
Sarah Hannett KC
Raj Desai
Roisin Swords-Kieley
(Instructed by Russell-Cooke LLP (Putney, London))

LORD HODGE, LADY ROSE AND LADY SIMLER (with whom Lord Reed and Lord Lloyd-Jones agree):

1. This appeal is concerned with establishing the correct interpretation of the Equality Act 2010 ("the EA 2010") which seeks to give statutory protection to people who are at risk of suffering from unlawful discrimination. The questions raised by this appeal directly affect women and members of the trans community. On the one hand, women have historically suffered from discrimination in our society and since 1975 have been given statutory protection against discrimination on the ground of sex. On the other hand, the trans community is both historically and currently a vulnerable community which Parliament has more recently sought to protect by statutory provision.

2. It is not the role of the court to adjudicate on the arguments in the public domain on the meaning of gender or sex, nor is it to define the meaning of the word "woman" other than when it is used in the provisions of the EA 2010. It has a more limited role which does not involve making policy. The principal question which the court addresses on this appeal is the meaning of the words which Parliament has used in the EA 2010 in legislating to protect women and members of the trans community against discrimination. Our task is to see if those words can bear a coherent and predictable meaning within the EA 2010 consistently with the Gender Recognition Act 2004 ("the GRA 2004").

3. As explained more fully below, the EA 2010 seeks to reduce inequality and to protect people with protected characteristics against discrimination. Among the people whom the EA 2010 recognises as having protected characteristics are women, whose protected characteristic is sex, and "transsexual" people, whose protected characteristic is gender reassignment.

4. The question for this court is a matter of statutory interpretation. But before discussing the general approach to statutory interpretation, we set out the structure of this judgment and address the matter of terminology.

5. We discuss terminology, the approach to statutory interpretation and the factual background between paras 6 and 35. We address the historical background to the GRA 2004, its interpretation and its operation between paras 36 and 111. We then between paras 112 and 264 address in some detail the interpretation of the EA 2010 to give its provisions a coherent and predictable meaning. We summarise our reasoning in para 265.

(1) Terminology

6. We are aware of the strength of feeling which has been generated by the disagreements between campaigners seeking to represent the interests of each of these groups and that taxonomy itself can generate controversy. We are content to draw on the terminology used by the Scottish Ministers in their written case for the purposes of this judgment and have adopted the following terms. A person who is a biological man, ie who was at birth of the male sex, but who has the protected characteristic of gender reassignment is described as a "trans woman". Similarly, a person who is a biological woman, ie who was at birth of the female sex, but who has the protected characteristic of gender reassignment is described as a "trans man". We describe trans women and trans men who have obtained a gender recognition certificate ("GRC") under the GRA 2004 as "trans women with a GRC" and "trans men with a GRC" respectively and their gender resulting from the GRC as their "acquired gender" or "acquired sex".

7. We also use the expression "biological sex" which is used widely, including in the judgments of the Court of Session, to describe the sex of a person at birth, and we use the expression "certificated sex" to describe the sex attained by the acquisition of a GRC.

(2) The question of statutory interpretation

8. The legislation with which this appeal is principally concerned is the EA 2010 and we address the effect, if any, of the GRA 2004 on the interpretation of the terms "sex", "man", "woman", and "male" and "female" used in the EA 2010. The central question on this appeal is whether the EA 2010 treats a trans woman with a GRC as a woman for all purposes within the scope of its provisions, or when that Act speaks of a "woman" and "sex" it is referring to a biological woman and biological sex.

9. The general approach to statutory interpretation in the United Kingdom is well-established. The House of Lords and this court have set out the basic approach on a number of occasions, including in *R v Secretary of State for the Environment, Transport and the Regions, Ex p Spath Holme Ltd* [2001] 2 AC 349. Most recently, this court set out the approach in *R (O) v Secretary of State for the Home Department* [2022] UKSC 3; [2023] AC 255 in which Lord Hodge DPSC, giving the leading judgment, stated (paras 29-31):

> "29. The courts in conducting statutory interpretation are 'seeking the meaning of the words which Parliament used': *Black-Clawson International Ltd v Papierwerke Waldhof-Aschaffenburg AG* [1975] AC 591, 613 per Lord Reid of Drem. More recently, Lord Nicholls of Birkenhead stated: 'Statutory interpretation is an exercise which requires the court to identify the meaning borne by the words in question in the particular context' (*R v Secretary of State for the Environment, Transport and the Regions, Ex p Spath Holme Ltd* [2001] 2 AC 349, 396). Words and passages in a statute derive their meaning from their context. A phrase or passage must be read in the context of the section as a whole and in the wider context of a relevant group of sections. Other provisions in a statute and the statute as a whole may provide the relevant context. They are the words which Parliament has chosen to enact as an expression of the purpose of the legislation and are therefore the primary source by which meaning is ascertained. There is an important constitutional reason for having regard primarily to the statutory context as Lord Nicholls explained in *Spath Holme*, 397: 'Citizens, with the assistance of their advisers, are intended to be able to understand parliamentary enactments, so that they can regulate their conduct accordingly. They should be able to rely upon what they read in an Act of Parliament.'
>
> 30. External aids to interpretation therefore must play a secondary role. Explanatory Notes, prepared under the authority of Parliament, may cast light on the meaning of particular statutory provisions. Other sources, such as Law Commission reports, reports of Royal Commissions and advisory committees, and Government White Papers may disclose the background to a statute and assist the court to identify not only the mischief which it addresses but also the purpose of the legislation, thereby assisting a purposive interpretation of a particular statutory provision. The context disclosed by such materials is relevant to assist the court to ascertain the meaning of the

statute, whether or not there is ambiguity and uncertainty, and indeed may reveal ambiguity or uncertainty: *Bennion, Bailey and Norbury on Statutory Interpretation*, 8th ed (2020), para 11.2. But none of these external aids displace the meanings conveyed by the words of a statute that, after consideration of that context, are clear and unambiguous and which do not produce absurdity. ...

31. Statutory interpretation involves an objective assessment of the meaning which a reasonable legislature as a body would be seeking to convey in using the statutory words which are being considered. ..."

10. In *R (Quintavalle) v Secretary of State for Health* [2003] UKHL 13; [2003] 2 AC 687, Lord Bingham of Cornhill warned against giving a literal interpretation to a particular statutory provision without regard to the context of the provision in the statute and the purpose of the statute. He stated (para 8):

"The court's task, within the permissible bounds of interpretation, is to give effect to Parliament's purpose. So the controversial provisions should be read in the context of the statute as a whole, and the statute as a whole should be read in the historical context of the situation which led to its enactment."

11. The general approach of focussing on the words which Parliament has used in a provision is justified by the principle that those are the words which Parliament has chosen to express the purpose of the legislation and by the expertise which the drafters of legislation bring to their task. But where there is sufficient doubt about the specific meaning of the words used which the court must resolve, the indicators of the legislature's purpose outside the provision in question, including the external aids described in para 30 of *R (O)* quoted above, must be given significant weight. As Lord Sales has stated in an extra-judicial writing, "sometimes the purpose for which legislative intervention was required may be the very prominent focus for the legislative activity which follows from it, and thus may frame in a particularly strong way the context in which that activity takes place" (see *The role of purpose in legislative interpretation: inescapable but problematic necessity*", Presentation at the Oxford University and University of Notre Dame Seminar on Public Law Theory: Topics in Legal Interpretation, 19 September 2024). Such aids can explain the meaning of a statutory provision which is open to doubt and can themselves alert the court to ambiguity in the provision, but they cannot displace the meanings conveyed by the clear and unambiguous words of a provision construed in the context of the statute as a whole.

12. Lord Nicholls' important constitutional insight in *Spath Holme*, that citizens with the help of their advisers should be able to understand statutes, points towards an interpretation that is clear and predictable. As Lord Hope DPSC stated in *Imperial Tobacco Ltd v Lord Advocate* [2012] UKSC 61; 2013 SC (UKSC) 153, at para 14:

"The best way of ensuring that a coherent, stable and workable outcome is achieved is to adopt an approach to the meaning of a statute that is constant and predictable. This will be achieved if the legislation is construed according to the ordinary meaning of the words used."

13. The presumption that a word has the same meaning throughout the Act when used

more than once in the same statute is consistent with this principle: see *Bennion, Bailey and Norbury on Statutory Interpretation*, 8th ed (2020) para 21.3. That presumption is based on the idea that the drafters of the statute were seeking to create a coherent statutory text. The weight to be given to the presumption depends upon the context in which the word or phrase appears in the instrument: *Assange v Swedish Prosecution Authority* [2012] UKSC 22; [2012] 2 AC 471, Lord Phillips of Worth Matravers PSC at para 75. The presumption may be stronger where a word is defined in the Act. In *R (Good Law Project) v Electoral Commission* [2018] EWHC 2414 (Admin), Leggatt LJ stated (para 33):

> "It is generally reasonable to assume that language has been used consistently by the legislature so that the same phrase when used in different places in a statute will bear the same meaning on each occasion – all the more so where the phrase has been expressly defined."

14. Whether Parliament intended a word to have a different meaning in different sections of an Act must be determined by looking at the context of the section in question and the Act as a whole.

(3) How the question arises

15. For Women Scotland ("the appellant") is a feminist voluntary organisation which campaigns to strengthen women's rights and children's rights in Scotland. This case is the second challenge by judicial review which the appellant has raised in relation to statutory guidance which the Scottish Ministers promulgated under section 7 of the Gender Representation on Public Boards (Scotland) Act 2018 ("the 2018 Act"). In the first petition for judicial review the appellant also asserted that the statutory definition of "woman" in the 2018 Act was outside the legislative competence of the Scottish Parliament under the Scotland Act 1998 as amended ("the Scotland Act"). Before we turn to the 2018 Act and the impugned statutory guidance, it may be helpful to outline the basis of that challenge under the Scotland Act.

16. Section 29 of the Scotland Act provides that a provision of an Act of the Scottish Parliament is outside the legislative competence of the Scottish Parliament if it relates to reserved matters. Schedule 5 to the Scotland Act specifies the matters which are reserved to the United Kingdom Parliament. One of the reserved matters (section L2) is "Equal opportunities". Since May 2016 there have been exceptions to the reservation of equal opportunities to allow the Scottish Parliament to legislate for positive action measures in relation to persons to be appointed to non-executive posts on the boards of certain public authorities in Scotland. Section L2 of Schedule 5 so far as relevant stated the exceptions as:

> "Equal opportunities so far as relating to the inclusion of persons with protected characteristics in non-executive posts on boards of Scottish public authorities with mixed functions or no reserved functions. Equal opportunities in relation to the Scottish functions of any Scottish public authority or cross-border public authority, other than any function that relates to the inclusion of persons in non-executive posts on boards of Scottish public authorities with mixed functions or no reserved functions. ..."

17. The Scottish Parliament passed the 2018 Act to provide for positive action measures to be taken in relation to the appointment of women to non-executive posts on boards of certain Scottish public authorities. The 2018 Act sets out a gender representation objective for a public board which is that "it has 50% of non-executive members who are women" (section 1(1)). The attainment of this objective is carefully circumscribed by section 4 which makes clear that preference can be given to a woman in order to further that objective only where there is no best candidate and only if the appointment of an equally qualified male candidate cannot be justified on the basis of his particular characteristics or situation. Section 2 of the 2018 Act defined "woman" as including:

> "a person who has the protected characteristic of gender reassignment (within the meaning of section 7 of the Equality Act 2010) if, and only if, the person is living as a woman and is proposing to undergo, is undergoing or has undergone a process (or part of a process) for the purpose of becoming female."

18. In its first judicial review the appellant challenged the statutory definition of "woman" in section 2 of the 2018 Act and paragraphs of the statutory guidance dated June 2020 which discussed that definition and explained that a trans woman had to meet the three criteria in section 2: to have the characteristic of gender reassignment, be living as a woman, and be proposing to undergo, be undergoing, or have undergone a process (or part of a process) as set out in the section 2 definition. The appellant was successful on appeal before the Second Division of the Inner House of the Court of Session (*For Women Scotland Ltd v Lord Advocate* [2022] CSIH 4; 2022 SC 150), which in para 40 of its judgment dated 18 February 2022 held that "transgender women" is not a protected characteristic under the EA 2010 and that the definition of "woman" adopted in the 2018 Act "impinges on the nature of protected characteristics which is a reserved matter". By interlocutor dated 22 March 2022 the Second Division declared that the definition of "woman" in section 2 of the 2018 Act was outside the legislative competence of the Scottish Parliament. In other words, because the definition of "woman" in section 2 of the 2018 Act included trans women as defined, it went beyond the scope of the exception permitted by section L2 of Schedule 5 to the Scotland Act; it therefore purported to legislate in respect of a reserved matter, namely equal opportunities, and so was outside the competence of the Scottish Parliament.

19. The response of the Scottish Ministers to this judicial decision was to issue fresh statutory guidance dated 19 April 2022. This guidance operated on the premise that the decision of the Second Division had nullified the definition of "woman" in section 2 of the 2018 Act. Instead, the Scottish Ministers asserted that a person who had been issued with a full GRC that her acquired gender was female, had the sex of a woman so that her appointment would count towards the achievement of the 50% objective. As explained below, this stance was consistent with the advice given by the Equality and Human Rights Commission ("EHRC"), which is the non-departmental public body in Great Britain with responsibility for promoting and enforcing equality and non-discrimination laws in England, Scotland and Wales.

20. The paragraph of the revised guidance which the appellant challenges states:

> "2.12 There is no definition of 'woman' set out in the Act with effect from 19 April 2022 following decisions of the Court of 18 February and 22 March

2022. Therefore 'woman' in the Act has the meaning under section 11 and section 212(1) of the Equality Act 2010. In addition, in terms of section 9(1) of the Gender Recognition Act 2004, where a full gender recognition certificate has been issued to a person that their acquired gender is female, the person's sex is that of a woman, and where a full gender recognition certificate has been issued to a person that their acquired gender is male, the person's sex becomes that of a man."

21. In July 2022 the appellant petitioned for judicial review to challenge the revised statutory guidance issued by the Scottish Ministers which it argues is unlawful because it is based on an error of law. The appellant seeks a declarator that the guidance is unlawful and an order for its reduction or the reduction of those parts which are found to be unlawful. The appellant argues that the guidance is not within the devolved competence of the Scottish Government under section 54 of the Scotland Act, which provides:

"(1) References in this Act to the exercise of a function being within or outside devolved competence are to be read in accordance with this section.

. . .

(3) In the case of any function other than a function of making, confirming or approving subordinate legislation, it is outside devolved competence to exercise the function (or exercise it in any way) so far as a provision of an Act of the Scottish Parliament conferring the function (or, as the case may be, conferring it so as to be exercisable in that way) would be outside the legislative competence of the Parliament."

22. The petition followed correspondence between the appellant's solicitors and the Scottish Government Legal Directorate ("SGLD"). In a letter dated 1 June 2022 to the appellant's solicitors the SGLD referred to the EHRC's guidance entitled "Separate and single-sex service-providers: a guide on the Equality Act sex and gender reassignment provisions" as updated in April 2022 in the light of the decision of the Inner House which we described above. The letter quoted from a section of the EHRC guidance, which was headed "What the Equality Act says about the protected characteristics of sex and gender reassignment" and which stated:

"Under the Equality Act 2010, 'sex' is understood as binary, being a man or a woman. For the purposes of the Act, a person's legal sex is their biological sex as recorded on their birth certificate. A trans person can change their legal sex by obtaining a Gender Recognition Certificate. A trans person who does not have a Gender Recognition Certificate retains the sex recorded on their birth certificate for the purposes of the Act."

The letter continued:

"This EHRC Guidance confirms that a trans woman with a full GRC has changed their legal sex from their biological sex (male) to their acquired sex (female). Therefore that trans woman has the protected characteristic under the 2010 Act of their acquired sex (female). In terms of the 2018 Act this means that a trans woman with a full GRC must be treated as a woman,

which is the position set out in the sentence in the Guidance on the 2018 Act
that your clients disagree with."

23. The Scottish Government's revised position therefore is that a trans woman with a
full GRC is treated by the EA 2010 as having the acquired sex of a woman and therefore
is a "woman" in sections 11 and 212(1) of the EA 2010. They accept that the wording
of the guidance set out in para 20 above is unfortunate in so far as it suggests that the
inclusion of trans women with a GRC is "in addition" to biological women included in
sections 11 and 212(1) of the EA 2010. On their case, therefore, the guidance would
mean exactly the same without the third sentence.

24. As explained more fully below, a person who is aged at least 18 can apply for a GRC
under the GRA 2004. Section 9(1) of that Act provides that when a full GRC is issued
to a person the person's gender becomes "for all purposes" the acquired gender so that
if the acquired gender is the female gender, the person's sex becomes that of a woman.
But that provision is "subject to provision made by this Act or any other enactment or
any subordinate legislation": section 9(3).

25. The central issue on this appeal is whether references in the EA 2010 to a person's
"sex" and to "woman" and "female" are to be interpreted in the light of section 9 of the
GRA 2004 as including persons who have an acquired gender through the possession of
a GRC.

26. The focus of this appeal is not on the status of the large majority of trans people who
do not possess a full GRC. Their sex remains in law their biological sex. This appeal
addresses the position of the small minority of trans people who possess a full GRC. Ben
Cooper KC, who appears for the intervener, Sex Matters, states in para 31 of his
written case that based on the most recent census data, the Office of National Statistics
estimated that there are about 48,000 trans men and 48,000 trans women in England
and Wales, and Scotland's census 2022 found that 19,990 people were trans, compared
with a total of 8,464 people who have ever obtained a GRC as at June 2024. He points
out that neither possession of a GRC nor the protected characteristic of gender
reassignment requires any specific physiological change.

(4) The decisions of the Court of Session

27. Lady Haldane heard the appellant's challenge in the Outer House. In a carefully
reasoned judgment dated 13 December 2022 ([2022] CSOH 90; 2023 SC 61), she
dismissed the petition. She rejected the appellant's argument that the Inner House's
decision in the first judicial review had authoritatively determined that "sex" in the EA
2010 was confined to biological sex only (para 44). She held that section 9(1) of the
GRA 2004 had the effect that a GRC changed a person's sex for all purposes, stating that
the language of section 9 of the GRA 2004 "could scarcely be clearer" (para 45). She
rejected the appellant's submission that the GRA 2004 had a narrow purpose which had
been largely superseded by subsequent legislation, including legislation establishing the
legality of same sex marriage. She observed that the GRA 2004 listed exceptions to the
rule in section 9(1), such as marriage, parenthood, succession, peerages and trusts, and
stated that the founding principle of section 9 of the GRA 2004 is a broad one: "that the
acquired gender becomes the person's sex 'for all purposes' subject to any other
enactments, or the statutory exceptions listed" (para 47). Lady Haldane rejected the

submissions (i) that there was a conflict between the GRA 2004 and the EA 2010, which she stated was "drafted in full awareness of the 2004 Act, and its ambit" (para 50), and (ii) that the EA 2010 impliedly repealed or disapplied section 9(1) of the GRA 2004 (para 52). As a result, "sex" in the EA 2010 was not confined to biological sex but includes the acquired sex of those who possess a GRC obtained under the GRA 2004. Lady Haldane therefore concluded that the revised guidance of the Scottish Ministers on the 2018 Act was lawful.

28. The Second Division of the Inner House (the Lord Justice Clerk (Lady Dorrian), Lord Malcolm and Lord Pentland) on 1 November 2023 refused the appellant's reclaiming motion ([2023] CSIH 37; 2023 SLT 1216). The Second Division, agreeing with Lady Haldane, held that the GRA 2004 was a far-reaching enactment which created a mechanism by which a person could change his or her sex in the eyes of the law. The judgment (para 42) stated that section 9(1), (2) and (3) of the GRA 2004 read together meant that a person with a GRC "acquires the opposite gender for all purposes unless there is a specific exception in the GRA [2004]; or unless the terms and context of a subsequent enactment require a different interpretation to follow". The judgment continued:

"Should that occur, however, it is to be expected that the inapplicability of section 9(1) would be clearly stated, or at the very least ... that the terms of the subsequent legislation are such that they are incompatible with, and would be rendered meaningless or unworkable by, the application of the general principle stated in section 9(1)."

29. The Second Division then examined the terms "sex" in sections 7 and 11 and "man" and "woman" in sections 11 and 212(1) of the EA 2010 and stated that such terms could have a biological meaning or could bear a wider meaning in accordance with the GRA 2004 so that a trans woman would be entitled to protection against discrimination on the ground of sex in her acquired gender as a woman. The terms "sex" and "gender" were often used interchangeably in the EA 2010. The provisions of the GRA 2004 and the EA 2010 could be interpreted consistently for the purposes of both statutes if the wider meaning were adopted. There was nothing in the EA 2010 that mandated a contrary conclusion. The Second Division then considered various provisions of the EA 2010 which the appellant argued were unworkable if the wider meaning of those words were adopted in section 11 of that Act. As we will be discussing those provisions below, it is sufficient at this stage to state that the provisions which the Second Division discussed related to: (i) the Armed Forces (Schedule 9), (ii) separate and single-sex spaces (section 29 and paragraphs 26 and 27 of Schedule 3), (iii) single-sex schools and institutions (Schedules 11 and 12), (iv) communal accommodation (paragraph 3 of Schedule 23), and (v) pregnancy and maternity (sections 4, 17 and 18). Of those provisions the Second Division held that only those relating to pregnancy and maternity might require a narrow interpretation of "woman" as meaning a biological woman. The Second Division also considered and rejected the submission that treating a trans woman as a woman under the EA 2010 would interfere with the right of freedom of association for lesbians. It concluded that persons with a GRC possess under section 11 the protected characteristic of sex according to the terms of their GRC as well as the protected characteristic of gender reassignment under section 7. The Second Division concluded that the Guidance on the 2018 Act was lawful because a person with a GRC in the female gender is a "woman" for the purposes of section 11 of the EA 2010.

30. The appellant now appeals to this court with the permission of the Second Division of the Inner House.

(5) The interventions in this appeal

31. Several persons and organisations applied to the court to intervene in this appeal. The court allowed four organisations to intervene in writing. Two of those four interveners were given permission to make oral submissions in addition to their written submissions.

32. First, the human rights charity, Sex Matters, whose object is to promote human rights where they relate to biological sex, in a focussed 20-page submission argues that "sex" in the EA 2010 should be construed as referring to biological sex principally because (i) trans women, including trans women with a GRC, are protected by the protected characteristic of gender reassignment, and (ii) the wider interpretation of the term "sex" in the EA 2010 leads to absurd or irrational results.

33. Secondly, the EHRC explains its longstanding view and policy position that the terms "sex", "man" and "woman" in the EA 2010 include those whose sex is certified in a GRC. The EHRC recognises that the wider definition which it favours causes difficulties and impairs the operation of the EA 2010 in four areas: (i) discrimination on the grounds of pregnancy and maternity (sections 17 and 18); (ii) the protection against sexual orientation discrimination (section 12(1)(a)) and in particular the risk that lesbians and gay men for whom the biological aspect of their same sex attraction is defining, might be precluded from forming associations which exclude trans women and trans men respectively; (iii) single-sex services (paragraphs 26-28 of Schedule 3), and (iv) communal accommodation (paragraph 3 of Schedule 23). The submission describes these difficulties as profound and suggests that Parliament should urgently resolve them.

34. The court also benefited from written interventions by Amnesty International UK, which submits that human rights principles demonstrate beyond doubt that the interpretation of the Scottish courts is correct. A combined written submission by Scottish Lesbians, the Lesbian Project and the LGB Alliance argues that a male can never be a lesbian as a matter of fact whether in possession of a GRC or not, and that the wider definition of "sex" and "woman" would create serious problems for lesbians in relation to services (section 29) and clubs and associations (sections 101 and 102 and Schedule 16) and would affect claims for direct and indirect discrimination (sections 13, 19 and 19A). The lesbian interveners also pray in aid of their submission rights under articles 8, 11 and 14 of the European Convention on Human Rights.

35. We are grateful to the interveners for their contributions. We are particularly grateful to Ben Cooper KC for his written and oral submissions on behalf of Sex Matters, which gave focus and structure to the argument that "sex", "man" and "woman" should be given a biological meaning, and who was able effectively to address the questions posed by members of the court in the hour he had to make his submissions.

(6) The legal background: the Sex Discrimination Act 1975

36. The Sex Discrimination Act 1975 ("the SDA 1975") came into force on 29 December 1975, on the same day as the Equal Pay Act 1970. The long title of the Act

described it as rendering unlawful certain kinds of sex discrimination and discrimination on the grounds of marriage. The structure of the SDA 1975 established the basis for the later legislation and several of the themes which are discussed later in this judgment emerge for the first time in this Act.

37. Section 1 of the SDA 1975 defined what amounted to discrimination against women. It provided that a person discriminates against a woman in any relevant circumstances if:

(a) on the ground of her sex he treats her less favourably than he treats or would treat a man: section 1(1)(a) (generally referred to as direct discrimination); or

(b) he applies to her a requirement or condition which he applies or would apply equally to a man but where the proportion of women who can comply with it is "considerably smaller" than the proportion of men who can comply: section 1(1)(b) (generally referred to as indirect discrimination).

38. Section 2(1) provided that section 1 and Parts 2 and 3 of the Act were to be read as applying equally to the treatment of men with such modifications to the wording as necessary. However, section 2(2) provided that in applying the Act to men "no account shall be taken of special treatment afforded to women in connection with pregnancy or childbirth".

39. Sections 5(2) and 82(1) of the SDA 1975 provided that in the Act "woman" includes a female of any age and "man" includes a male of any age. Similarly, in the Equal Pay Act 1970, section 11(2) provided that "In this Act the expressions 'man' and 'woman' shall be read as applying to persons of whatever age".

40. Part 2 of the SDA 1975 dealt with discrimination in the employment field. Section 6(1) made it unlawful for a person to discriminate against a woman in relation to the arrangements he makes for choosing who should be offered a job, in the terms on which he offers her the job or by refusing to offer her the job. Section 6(2) made it unlawful to discriminate against an employed woman in the way that access to opportunities for promotion, training or other services were offered or by dismissing her. There were several exceptions to the prohibition in section 6 which were designed to establish the boundary between the SDA 1975 and the Equal Pay Act 1970. Broadly, subsections (4) to (7) of section 6 excepted discrimination as regards pay and pensions from this prohibition on the basis that differential treatment of this kind would be dealt with under the Equal Pay Act 1970.

41. Certain exceptions were built into the legislation, some of which were repealed long before the whole Act was superseded by the EA 2010. For example, according to section 6(3) as originally enacted, the prohibition on discrimination under section 6(1) and (2) did not apply to employment "for the purposes of a private household" or where the number of people employed was not more than five. The exception for small employers was repealed by the Sex Discrimination Act 1986 and the private household exception re-enacted in a much narrower form by section 1(2) of the 1986 Act, limiting it to where objection might reasonably be taken by a person living in the home to physical or social contact with someone of the opposite sex.

42. Section 7 of the SDA 1975 as enacted provided the exception which is reflected in the subsequent legislation, namely that discrimination is not unlawful where sex is a genuine occupational qualification ("GOQ"). The exception does not apply to discrimination in the terms and conditions on which a woman is employed; once a woman has been engaged in the job, there can be no genuine occupational reason for giving her less favourable terms and conditions than her male colleagues. The circumstances in which the defence of GOQ could be relied upon included the following:

(a) Where the essential nature of the job called for a man for reasons of physiology (other than physical strength or stamina), or for reasons of authenticity in dramatic performances: section 7(2)(a);

(b) Where the job needed to be held by a man to preserve decency or privacy because it was likely to involve physical contact or where men would be in a state of undress or using sanitary facilities: section 7(2)(b);

(c) Where the job holder had to live in premises provided by the employer and there were no facilities to accommodate women either to sleep separately or to use sanitary facilities. This was subject to the proviso that the exception applied only if it was not reasonable to expect the employer to provide separate facilities: section 7(2)(c);

(d) The job holder worked in a prison or hospital where all the people present were men and it was reasonable that the job should not be held by a woman: section 7(2)(d).

43. The defence of a GOQ could be relied on where only some of the duties of the job fell within the circumstances described but it could not be relied on in respect of a vacancy where the employer already had enough male employees to carry out those duties: section 7(3) and (4).

44. The SDA 1975 exempted a range of jobs from the ambit of the Act in whole or in part. For example, as regards prison officers it was not unlawful to impose a height requirement on both male and female prison officers: see section 18(1). Further the Act made some textual amendments to earlier legislation which assumed that all employees in occupations covered by that legislation would be men. For example, the provision in the Mines and Quarries Act 1954 which provided that no female should be employed below ground at a mine was modified to apply only to jobs where the duties ordinarily required the employee to spend a significant proportion of his time below ground: see section 21(1) of the SDA 1975. The language used in the Coal Mines Regulation Act 1908 was also modified to reflect the fact that women might now be employed; for the words "workman" or "man" there were substituted "worker": section 21(2).

45. Part 3 of the SDA 1975 dealt with discrimination in fields other than employment, in particular schools and universities (with an exception for single-sex establishments) and in the provision of goods, facilities or services. Section 29 provided that it was unlawful to discriminate on grounds of sex in the provision of a wide range of services including banking, transport, recreation and the services of any trade or local authority.

46. Again, there were various exceptions such as providing accommodation where the provider intended to continue to reside at the premises: section 32(1)(a). Section 35(1) provided a more general exception to the prohibition in section 29(1) for a person who provided facilities or services restricted to men where, for example (section 35(1)(c)):

> "(c) the facilities or services are provided for, or are likely to be used by, two or more persons at the same time, and
>
> (i) the facilities or services are such, or those persons are such, that male users are likely to suffer serious embarrassment at the presence of a woman, or
>
> (ii) the facilities or services are such that a user is likely to be in a state of undress and a male user might reasonably object to the presence of a female user."

47. Further, there was an exception where it was likely that there would be physical contact between the user of the facilities and another person and that other person might reasonably object if the user was a woman: see section 35(2).

48. Part 5 of the SDA 1975 conferred further general exceptions. These included the following:

> (a) Section 44 provided that nothing prevented excluding men from women's sporting competitions or other activities of a competitive nature where the physical strength, stamina or physique of the average woman put her at a disadvantage to the average man.
>
> (b) Section 46 made further provision about maintaining single-sex communal accommodation provided that the accommodation was managed in a way which "comes as near as may be to fair and equitable treatment of men and women".
>
> (c) Section 49 provided for ensuring appropriate representation on the bodies of trade unions, employer organisations and other professional or trade bodies. Where the body concerned was made up wholly or mainly of elected members it would not be unlawful to reserve seats on the body for persons of one sex in order to ensure that a minimum number of persons of that sex were members, if this was needed "to secure a reasonable lower limit to the number of members of that sex serving on the body".

49. The SDA 1975 was amended in important respects before being repealed by the EA 2010. In 2005 and 2008, provisions were inserted by the Employment Equality (Sex Discrimination) Regulations 2005 (SI 2005/2467) and the Sex Discrimination (Amendment of Legislation) Regulations 2008 (SI 2008/963) to prohibit discrimination against women on the ground of pregnancy or maternity leave both in employment (section 3A) and in the provision of services etc (section 3B).

50. What we draw from this consideration of the SDA 1975 are the following points.

51. First, there can be no doubt that Parliament intended that the words "man" and

"woman" in the SDA 1975 would refer to biological sex – the trans community of course existed at the time but their recognition and protection did not.

52. Secondly, the legislation recognised and accommodated the reasonable expectations of people that in situations where there was physical contact between people, or where people would be undressing together or living in the same premises, or using sanitary facilities together, considerations of privacy and decency required that separate facilities be permitted for men and women.

53. Thirdly, a range of other exceptions were considered necessary and reasonable, particularly (a) in relation to sport and competitive activity where typical masculine physique would give an unfair advantage and (b) where positive action was needed to ensure that there was a reasonable representation of men and women on the boards of certain bodies.

(7) Discrimination on the grounds of being transgender: the 1999 Regulations

54. The common law of England and Wales did not recognise the possibility of a person becoming a different gender from their gender at birth. In the well-known case of *Corbett v Corbett (otherwise Ashley)* [1971] P 83, the High Court declared that a marriage was null and void where both parties were biological males but one had undergone gender reassignment. Ormrod J said that over a very large area, the law is indifferent to sex. In other areas, such as insurance and pension schemes, there was nothing to prevent the parties to a contract from agreeing that the person concerned should be treated as a man or a woman, as the case may be: p 105. But marriage was a relationship between a man and a woman and, in the context of marriage, even if not for other purposes, the person was still a biological male. That conclusion that a person could not change sex was applied in the criminal law in *R v Tan* [1983] QB 1053.

55. In *P v S and Cornwall County Council* (Case C-13/94) [1996] ICR 795, [1996] ECR I-2143 ("*P v S*") the European Court of Justice considered the scope of the Equal Treatment Directive, that is Council Directive 76/207/EEC (OJ 1976 L39 p 40) in the context of alleged discrimination connected to gender reassignment. The applicant (a biological male employee) was dismissed by Cornwall County Council after telling her employer that she intended to undergo gender reassignment surgery. She complained of unlawful discrimination on the grounds of her sex. The Judge Rapporteur recorded that the industrial tribunal "found that there was no remedy under the Sex Discrimination Act 1975, the applicable United Kingdom statute, since English law took cognisance only of situations in which men or women were treated differently because they belonged to one sex or the other, and did not recognise a transsexual condition in addition to the two sexes. Under English law, the applicant was at all times a male" (para 7). The Court at para 18 held that the Directive was "simply the expression, in the relevant field, of the principle of equality, which is one of the fundamental principles of Community law". The right not to be discriminated against on grounds of sex was, the Court said, a fundamental human right and accordingly the Directive also applied to discrimination arising from gender reassignment (para 20).

56. The *P v S* decision led to the adoption of the Sex Discrimination (Gender Reassignment) Regulations 1999 (SI 1999/1102) ("the 1999 Regulations"). The 1999 Regulations amended the SDA 1975 in important ways.

57. First, regulation 2 inserted section 2A which defined discrimination as including treating a person, B, less favourably "on the ground that B intends to undergo, is undergoing or has undergone gender reassignment" for the purposes of any provision in Part 2 or, subject to a limited exception, Part 3 of the SDA 1975. A definition of "gender reassignment" was inserted into section 82 of the SDA 1975:

> "'gender reassignment' means a process which is undertaken under medical supervision for the purpose of reassigning a person's sex by changing physiological or other characteristics of sex, and includes any part of such a process ..."

58. The 1999 Regulations did not insert a free-standing prohibition on discrimination separate from section 6. Rather, the prohibition on discriminating against a woman now prohibited direct discrimination as defined by section 2A, namely on the grounds of gender reassignment, but only in the employment field. It was therefore unlawful under section 6 for A to discriminate against a woman in the ways caught by section 6 on the ground that she intended to undergo or was undergoing or had undergone gender reassignment. In light of section 2 of the SDA 1975, this also made it unlawful under section 6 for A to discriminate against a man if A treated him less favourably on that ground. However, the subsections of section 6 which prevented the overlap with the Equal Pay Act 1970 were disapplied so that discrimination in respect of pay and pensions on the grounds of gender reassignment was prohibited under section 6: see the new section 6(8) inserted by regulation 3(1) of the 1999 Regulations.

59. Regulation 4 of the 1999 Regulations inserted section 7A which provided for an exception to the prohibition of discrimination in section 6(1) and (2) of the SDA 1975 where the discrimination fell within section 2A but where "being a man" or "being a woman" was a GOQ for the job and the treatment was reasonable in view of the circumstances described in section 7(2) and any other relevant circumstances.

60. Further, section 7B was inserted into the SDA 1975 to provide an additional exception to the unlawfulness of discrimination under certain elements of section 6(1) where there was a "supplementary genuine occupational qualification" for the job. A supplementary GOQ was defined in the new section 7B(2) as arising only in the circumstances set out in subsection (2). Thus:

> (a) The holder of the job was "liable to be called upon to perform intimate physical searches pursuant to statutory powers": section 7B(2)(a).

> (b) The holder of the job had to live in a private home and the job involved a degree of physical or social contact with a person living in the home or knowledge of the intimate details of that person's life and that person might reasonably object to the job being held by someone who was undergoing or who had undergone gender reassignment: section 7B(2)(b).

> (c) The holder of the job would have to share accommodation provided by the employer with other employees who, for the purpose of preserving decency and privacy, might reasonably object to sharing the accommodation and facilities with someone whilst the job holder was undergoing gender reassignment: section 7B(2)(c) and 7B(3).

(d) The holder of the job provided personal services to vulnerable individuals and the employer's reasonable view was that the services could not be effectively provided by someone undergoing gender reassignment: section 7B(2)(d) and 7B(3).

61. Some of these exceptions (such as that described in (b) above) were limited to where the person was undergoing or had undergone gender reassignment and did not except discrimination where the person intended to undergo gender reassignment. Others (such as that described in (c) above) applied only where the person intended to undergo or was undergoing gender reassignment but not where the person had undergone gender reassignment. Similar exceptions to discrimination were also provided for other forms of employment, including contract workers (regulation 4(2)-(3) amending section 9 of the SDA 1975), and partnerships (regulation 4(4)-(5) amending section 11 of the SDA 1975).

62. The 1999 Regulations did not amend the definitions of "man" and "woman" in the SDA 1975.

(8) The GRA 2004 as enacted

63. The enactment of the GRA 2004 was prompted by the judgment of the European Court of Human Rights ("ECtHR") in *Goodwin v United Kingdom* (Application No 28957/95) (2002) 35 EHRR 18 (*"Goodwin"*) and by a declaration of incompatibility made by the House of Lords in *Bellinger v Bellinger* [2003] UKHL 21, [2003] 2 AC 467 (*"Bellinger"*). In *Goodwin*, the applicant's biological sex was male but she had undergone gender reassignment surgery. The ECtHR held that it was a breach of the applicant's right to respect for private life under article 8 of the Convention for there to be no legal recognition of her acquired gender. The ECtHR described the applicant as having initially undergone hormone therapy, grooming classes and voice training and as having "lived fully as a woman" since 1985. She later underwent gender reassignment surgery at a National Health Service hospital. The Court referred to various difficulties faced by the applicant because of the failure of the law to recognise her acquired gender. These included her inability to change her birth certificate, and different treatment as regards social security and national insurance issues, pensions and employment. The Court recognised that it had previously held that UK law did not interfere with respect for private life: para 73. But in the light of the then social conditions, it reassessed the appropriate application of the Convention.

64. The ECtHR was struck in particular by the fact that the National Health Service recognised the condition of gender dysphoria and provided reassignment surgery "with a view to achieving as one of its principal purposes as close an assimilation as possible to the gender in which the transsexual perceives that he or she properly belongs" (para 78). Yet there was no legal recognition of her changed status in law. The Court discussed medical evidence about the causes of what it called "transsexualism" and noted that the vast majority of Contracting States, including the UK, provided treatment including irreversible surgery. However, the ongoing debate about the exact causes of the condition were of diminished relevance because "given the numerous and painful interventions involved in such surgery and the level of commitment and conviction required to achieve a change in social gender role" it could not be suggested that there was "anything arbitrary or capricious in the decision taken by a person to undergo

gender re-assignment": para 81.

65. The Court concluded that the unsatisfactory situation in which post-operative transsexuals live in an intermediate zone which is not quite one gender or the other was no longer sustainable: para 90.

66. The *Goodwin* judgment was considered by the House of Lords in *Bellinger* where their Lordships were invited to declare a marriage valid which had been entered into by a man and a trans woman. Their Lordships declined to do so. Lord Nicholls referred to *Goodwin* and the Government's announcement that it intended to bring forward primary legislation to address the issue. He said that recognition of Mrs Bellinger as female for the purposes of section 11(c) of the Matrimonial Causes Act 1973 "would necessitate giving the expressions 'male' and 'female' in that Act a novel, extended meaning: that a person may be born with one sex but later become, or become regarded as, a person of the opposite sex": para 36. Lord Nicholls went on:

> "37. This would represent a major change in the law, having far reaching ramifications. It raises issues whose solution calls for extensive enquiry and the widest public consultation and discussion. Questions of social policy and administrative feasibility arise at several points, and their interaction has to be evaluated and balanced. The issues are altogether ill-suited for determination by courts and court procedures. They are pre-eminently a matter for Parliament, the more especially when the Government, in unequivocal terms, has already announced its intention to introduce comprehensive primary legislation on this difficult and sensitive subject."

67. The House of Lords held further that it was not possible to "read down" the 1973 Act and made a declaration of incompatibility under section 4 of the Human Rights Act 1998.

68. The GRA 2004 came into force on 4 April 2005 and provides a framework for recognising a person's reassigned gender. The compatibility of the UK's provision for recognition of gender reassignment with article 8 of the Convention was considered by the ECtHR again in *Grant v United Kingdom* (Application No 32570/03) (2006) 44 EHRR 1. There a trans woman complained that she was only entitled to receive her state pension at age 65, the age for men, rather than at 60, the age for women. She had been issued with a GRC once the GRA 2004 came into force. The Court held that the duration of the applicant's victim status lasted from the occasion on which she was refused a pension following the Court's judgment in *Goodwin* until the passing of the GRA 2004: para 43.

69. The main provisions of the GRA 2004:

(a) provided for applications to be made for a GRC and for the criteria to be applied and the evidence to be provided: sections 1, 2 and 3;

(b) established a Gender Recognition Panel ("the Panel") to determine those applications and provided for appeals from decisions of the Panel: section 1(3) and Schedule 1;

(c) provided for the consequences of the issue of a gender recognition certificate, including the creation and maintenance of the Gender Recognition Register described in Schedule 3;

(d) provided for a prohibition on disclosure of protected information about a person who has made an application: section 22;
(e) provided for limited amendments to the SDA 1975.

70. We discuss each of these briefly in turn, focussing for present purposes on the text of the GRA 2004 as originally enacted, since neither party has suggested that any of the amendments made to the GRA 2004 can affect how it applies to the EA 2010.

(i) Applications for gender recognition certificates

71. Section 1 of the GRA 2004 provides that a person aged 18 or over can apply for a GRC on the basis of "living in the other gender" or having changed gender in an overseas country. Section 2 provides that where the application is based on the person living in the other gender, the Panel must grant the application if satisfied that the applicant satisfies four criteria, namely that the applicant:

(a) has or has had gender dysphoria,

(b) has lived in the acquired gender throughout the period of two years ending with the date on which the application is made,

(c) intends to continue to live in the acquired gender until death, and

(d) complies with the evidential requirements imposed by and under section 3.

72. The evidence required under section 3 includes two medical reports, one of which must be by a registered medical practitioner or chartered psychologist practising in the field of gender dysphoria, and that report must include "details of the diagnosis of the applicant's gender dysphoria". Further, if the applicant has undergone or is undergoing or plans to undergo treatment to modify sexual characteristics, one of the reports must include details of that treatment. The applicant must also provide a statutory declaration that the applicant has lived in the acquired gender for two years and intends to do so until death.

73. From its enactment, the GRA 2004 went further than the decision in *Goodwin* may strictly have required at that point to ensure compliance with article 8. The applicant in *Goodwin* had undergone what the ECtHR described as "the long and difficult process of transformation" (para 78), but the GRA 2004 recognised a broader class of transgender people as entitled to formal recognition even if they had not undergone surgery. In that respect, the GRA 2004 anticipated the decision of the ECtHR in *AP, Garçon and Nicot v France* (Applications Nos 79885/12, 52471/13 and 52596/13, judgment of 6 April 2017). In that case the Court held that it was a breach of article 8 to make legal recognition of a person's transgender status conditional on sterilisation surgery or on treatment which entailed a very high probability of sterility: see para 120 of the judgment. The Court noted that imposing such a pre-condition presented

transgender persons "with an impossible dilemma" if they did not want to undergo sterilisation surgery or treatment. That condition amounted to a violation of article 8. However, there was no breach of article 8 in requiring a diagnosis of gender dysphoria. There was at that time near-unanimity amongst Contracting States in requiring such a diagnosis and imposing that requirement did not infringe article 8: see para 140.

74. Applications for a GRC are determined by the Panel in private and, according to section 4 of the GRA 2004, if the Panel grants the application it must issue a GRC to the applicant. The certificate is either a full certificate if the applicant is not married or an interim certificate if the applicant is married. The Act contains complex provisions for addressing the issues raised by the response of the applicant's spouse to the successful application: see Schedule 4 to the Act. The issue of an interim certificate is a ground for divorce and if divorce ensues, the applicant must then be granted a full GRC. Appeals on a point of law from the rejection of an application go to the High Court or Court of Session: section 8. The certificate must state that the acquired gender is male or is female: the Panel has no power to issue a "non-binary" certificate, even where the applicant has a certificate declaring them to be "non-binary" issued by an overseas authority: see *R (Castellucci) v Gender Recognition Panel* [2024] EWHC 54 (Admin), [2024] KB 995.

75. Section 9 of the GRA 2004 is key to the issues raised in this appeal. It remains in force in the form originally enacted and provides:

> "9 General
>
> (1) Where a full gender recognition certificate is issued to a person, the person's gender becomes for all purposes the acquired gender (so that, if the acquired gender is the male gender, the person's sex becomes that of a man and, if it is the female gender, the person's sex becomes that of a woman).
>
> (2) Subsection (1) does not affect things done, or events occurring, before the certificate is issued; but it does operate for the interpretation of enactments passed, and instruments and other documents made, before the certificate is issued (as well as those passed or made afterwards).
>
> (3) Subsection (1) is subject to provision made by this Act or any other enactment or any subordinate legislation."

76. Section 10 and Schedule 3 address the effect of the GRC on the UK birth register entry in relation to the recipient. Other Schedules to the Act deal with amendments to marriage law (Schedule 4) and entitlements to social security benefits and pensions (Schedule 5).

77. The GRA 2004 also expressly excepted particular matters, providing that they were not to be affected wholly or in part by the grant of the GRC. Succession, the descent of peerages, the administration of trusts and the disposition of property under a will were effectively excepted from the regime by sections 15 to 18. Other exceptions, some of which remain in force, were provided for as follows:

(a) Section 12 provided that the fact that a person's gender has become the acquired gender does not affect the status of the person as the father or mother of a child.

(b) Section 19 provided that a person may be excluded from participating as a competitor in a "gender-affected sport" if that was necessary to secure fair competition or the safety of competitors. A "gender-affected sport" was defined as one where "the physical strength, stamina or physique of average persons of one gender would put them at a disadvantage to average persons of the other gender as competitors in events involving the sport."

(c) Section 20 provided that the receipt of a GRC did not prevent a person from being convicted of a gender-specific offence which can be committed only by a person of their biological gender or from being a victim of an offence of which only people of their biological gender can be victims.

78. Section 22 provides for the confidentiality of "protected information" and remains in force. It is an offence for a person who has acquired protected information in an official capacity to disclose that information to any other person. "Protected information" means information which concerns an application for a GRC or, if a GRC had been issued, concerned the person's gender before the acquired gender. Subsection (4) provides gateways for the lawful disclosure of protected information, including where the information does not enable the applicant to be identified, or where the person has agreed to the disclosure or for certain other purposes, including circumstances to be prescribed by the Secretary of State.

79. Section 23 conferred on the Secretary of State and on Scottish Ministers and the appropriate Northern Ireland department a general power to make orders, following appropriate consultation, modifying the operation of any enactment or subordinate legislation in relation to persons whose gender has become the acquired gender. This power was subsequently used to modify Scottish laws on marriage.

80. Schedule 6 to the GRA 2004 made amendments to the SDA 1975, in particular amendments to sections 7A and 7B (inserted by the 1999 Regulations). Neither Schedule 6 nor any other provision in the GRA 2004 made any express amendment to the definition of "man" and "woman" in the SDA 1975.

81. As to what one can glean from the provisions of the GRA 2004 about the intended effect of section 9(1) on the scope of the SDA 1975, the Scottish Ministers drew the court's attention to para 27 of the Explanatory Notes. The notes give as an example of the effect of section 9(1), that a trans man with a GRC would be entitled to protection from discrimination as a woman under the SDA 1975. In our view, this is a good illustration of why the use to which the courts should put explanatory notes is limited to the context of the legislation and the mischief to which its provisions are aimed: see Lord Steyn in R (Westminster City Council) v National Asylum Support Service [2002] UKHL 38, [2002] 1 WLR 2956, para 5 and the passages from R (O) cited earlier. There is nothing in the notes to suggest that the department had undertaken the kind of detailed analysis of the effect of such a change on the operation of provisions of the SDA 1975, as amended by the 1999 Regulations, that we have undertaken in the following sections of this judgment before giving that as an example of the effect of section 9(1).

82. The Scottish Ministers make a different point on the scope of the amendments made to the SDA 1975 by Schedule 6 to the GRA 2004. The amendments made to sections 7A and 7B disapplied the exceptions for GOQs and supplementary GOQs in those sections if the discrimination was against a person whose gender had become the acquired gender under the GRA 2004. The effect of the amendments made by Schedule 6 is to add a provision removing the exception – so discrimination is not permitted – where the discrimination is against a person whose gender has become the acquired gender under the GRA 2004. In our judgment these provisions say nothing about the intended effect more generally of section 9(1) on the meaning of the terms "man" and "woman" in the SDA 1975. In any event, the provisions regarding GOQs in sections 7A and 7B as amended were not carried forward into the EA 2010. Schedule 9 to that Act made fresh provision for GOQs.

(ii) Guidance and case law on applications for gender recognition certificates

83. The criteria in section 2 and the evidence requirement in section 3 of the GRA 2004 have been the subject of guidance and some case law. There are several authorities which describe the condition of gender dysphoria which must be diagnosed before the applicant can apply for a GRC. In *R (C) v Secretary of State for Work and Pensions* [2017] UKSC 72, [2017] 1 WLR 4127 ("*R (C) v DWP*"), Lady Hale PSC described gender dysphoria as "the overwhelming sense that one has been born into the wrong body, with the wrong anatomy and the wrong physiology": para 1. She referred also to the transgender person's "deep need to live successfully and peacefully in their reassigned gender, something which non-transgender people can take for granted".

84. So far as the medical reports required by section 3 are concerned, the President of the Panel issued guidance in 2005 pursuant to paragraph 6(5) of Schedule 1 (as amended) as to how panels should consider the medical evidence. The guidance states:

> "3. ... the Panel must therefore examine the medical evidence provided in order to determine whether it is satisfied that the applicant has or has had the diagnosis of gender dysphoria. In order to do so the Panel requires more than a simple statement that such a diagnosis was made. The medical practitioner practising in the field who supplies the report should include details of the process followed and evidence considered over a period of time to make the diagnosis in the applicant's case. Nor is it sufficient to use the broad phrase 'gender reassignment surgery' without indicating what surgery has been carried out. Nor should relevant treatments be omitted, such as hormone therapy. These requirements are particularly pertinent in assisting the Panel to be satisfied not only that the applicant has or has had gender dysphoria but also has lived in the acquired gender for at least two years and intends to live in that gender until death.
>
> 4. On the other hand, doctors need not set out every detail which has led them to make the diagnosis. What the Panel needs is sufficient detail to satisfy itself that the diagnosis is soundly based and that the treatment received or planned is consistent with and supports that diagnosis."

85. In *Carpenter v Secretary of State for Justice* [2015] EWHC 464 (Admin), [2015] 1 WLR 4111 a trans woman challenged the requirement under section 3(3) of the GRA 2004

that she had to provide details to the Panel of the surgical treatment she had undergone for the purpose of modifying her sexual characteristics. She argued this infringed her article 8 rights because applicants could be issued with a certificate without having undergone surgery and without therefore having to provide such details. The challenge was rejected. Thirlwall J accepted that the requirement to provide medical details engaged the article 8 right to respect for private life. However, where an applicant had undergone surgery, or planned to do so, that fact was highly relevant, if not central, to the application and was plainly necessary to the Panel's consideration of the criteria in section 2(1)(a) to (c) of the GRA 2004. Thirlwall J said at para 23:

> "Undergoing or intending to undergo surgery for the purposes of modifying sexual characteristics is overwhelming evidence of the existence now or previously of gender dysphoria and of the desire of the applicant to live in the acquired gender until death. No competent, conscientious medical practitioner could produce a report on gender dysphoria (past or present) which did not refer to treatment received."

86. She also recorded at para 24 of her judgment that counsel for the Secretary of State had told the court that where an applicant has not undergone any treatment, it is the Panel's usual procedure to require the second report submitted by the applicant to explain why this is the case. She concluded (para 28) that given that the information was necessary to the decision to be taken and that its dissemination beyond the Panel was prohibited, the provision of the information was necessary and proportionate to the legitimate aim and that there was no breach of article 8.

(iii) Living in an acquired gender

87. Many of the judgments handed down in earlier cases addressing transgender issues emphasise the importance to the trans person who had brought the proceedings before the court of modifying their appearance so that they look like a typical person of their acquired gender. For example, *Chief Constable of the West Yorkshire Police v A (No 2)* [2004] UKHL 21, [2005] 1 AC 51 concerned a claim under sections 1 and 6 of the SDA 1975 by a trans woman prior to the coming into force of the GRA 2004. The issue was whether she could be refused appointment as a police officer because in the chief constable's view she was not able to carry out intimate searches of either men or women. She could not search men because she presented as a woman and she could not search women because she was male as a matter of law. In her speech, Lady Hale said at para 61 that the applicant "has done everything that she possibly could do to align her physical identity with her psychological identity. She has lived successfully as a woman for many years. She has taken the appropriate hormone treatment and concluded a programme of surgery. She believes that she presents as a woman in every respect". Similarly, in *R (C) v DWP* Lady Hale PSC recorded that the applicant in the proceedings before the court had undergone full gender reassignment treatment and surgery which included facial feminisation surgery "in [the applicant's] words because it was 'incredibly important' to her 'easily to "pass" as a woman.'": para 3.

88. However, the requirements in section 2(1)(b) and (c) of the GRA 2004 have not been interpreted to require, for example, biological men to prove that they have modified or intend to modify their physical appearance so as to "pass" as a woman in order to establish that they have been "living as" women in the past and that they intend

to do so until death.

89. The court was provided with guidance on completing the application form for a GRC issued by His Majesty's Courts and Tribunals Service (rather than by the President of the Panel) (Reference T451). Section 5 of the guidance deals with "Time living in your acquired gender". Applicants must enter the date from which they can prove that they have been living full time in their acquired gender. The evidence that the guidance suggests that the applicant provide is in the form of documents that are dated and include the applicant's name in the acquired gender. Examples of the documents that can be used are driving licences and passports, payslips, bank statements, official letters from doctors or dentists, utility bills or academic certificates. The guidance states that typically five or six different documents should be included.

90. The court was not provided with any further explanation of what names are regarded as being in any particular gender or whether this refers only to the pronouns used. The guidance also refers to the making of the statutory declaration that the applicant has lived as a male or female and intends to live in that gender until death. There is no guidance as to what it means to live in a gender, other than to ensure that the person's name in certain documents is a name in the acquired gender.

91. The application of this guidance and the relationship between the criteria in sections 2 and 3 were considered by the Divisional Court (Sir Andrew McFarlane P and Lieven J) in *AB v Gender Recognition Panel* [2024] EWHC 1456 (Fam), [2025] 1 WLR 227. There the appellant appealed against the Panel's decision to refuse her application for a GRC in the female gender. The judgment records at paras 7 and 8 that the applicant had provided a range of documents showing that she had changed her name. It noted also that the medical report from a doctor practising in the field of gender dysphoria had described the applicant's goals as regards treatment as requiring "a basic biological incompatibility" since she wanted both to achieve a gynaecoid body shape including adult female breast development and also to retain the capacity "to have a functional penis, with capacity for erection and genital sexual response": para 27. It was, the doctor said, for the applicant "to decide what takes priority". The second medical report provided with AB's application recorded that at interview, the applicant had presented as "straightforwardly feminine" and that she had "moved into a stable female social role".

92. In her judicial review, she challenged the Panel's conclusion that there was very little evidence that the applicant was "living in real life as a female". She submitted that she had followed the guidance by providing as evidence her passport (stating her sex as "F"), deed polls by which she had adopted female names and bank statements with her female name prefixed by "Miss": para 42. She also referred to the hormone treatment and testosterone blocking medication that she had taken except for a short period. The Court criticised the Panel's decision letter for failing to analyse properly the evidence supporting the applicant's assertion that she had been "living in the acquired gender". The decision had not referred to the passport, deed polls or bank account statements and had not given any reasons as to why they were dissatisfied with this evidence. The Court held that the Panel had erred in considering only the medical evidence on the question of whether she had been living as a woman: para 62. That was only one part of the evidence before the Panel on the issue and some of the statements in the medical reports were supportive of her assertion that she had been living in the female gender.

93. The Court concluded at para 67 that the evidence taken as a whole presented a clear and consistent picture of a person who had lived as a female. Although she had for a limited time some years before stopped hormone treatment in order to retain some male sexual function, all the other material, including important official documents, indicated a consistent course of conduct in living her life as female. The Court recorded that further evidence had been provided to enable the Court to make a decision whether to issue her with a GRC in the event that the Panel's decision was set aside. This included a statement that "she has lived as a female since 2012 to the extent that she believes that many of her friends and acquaintances would not know that she had been born male": para 76. The Court issued a GRC. The Court did not therefore address the question whether, if AB's evidence showed that she complied with the guidance because her official documents used female names and pronouns but that she did not "present" as female or occupy a stable female social role, she would have satisfied the criteria that she was living as a woman and that she intended to continue to do so.

(iv) Case law on the effect of section 9(1)

94. As explained earlier, the principal issue in this appeal is the effect of section 9 of the GRA 2004 on the meaning of the words "man" and "woman" in the EA 2010. Section 9 (set out at para 75 above) provides both for a rule that on receipt of a GRC "the person's gender becomes for all purposes the acquired gender" (subsection (1)) and also a carve out from the operation of that rule, namely that it is subject to a provision made in the GRA 2004 itself or in any other enactment or any subordinate legislation (subsection (3)).

95. In her submissions on this point, Ms Crawford KC on behalf of the Scottish Ministers compared section 9(1) with section 40 of the Adoption and Children (Scotland) Act 2007. That provides that an adopted person is to be treated in law as if born as the child of the adopters or adopter (section 40(1)) and further is to be treated in law "as not being the child of any person other than the adopters or adopter" (section 40(3) and (4)). However, that deeming provision is not as absolute as the wording may suggest since subsection (7) provides that the court may direct that subsection (4) does not apply or applies only to the extent specified in the direction.

96. Both parties referred to the case of *Fowler v Revenue and Customs Commissioners* [2020] UKSC 22, [2020] 1 WLR 2227 on the court's approach to the similar issue of interpreting and applying statutory deeming provisions. The court recognises that it would be entirely incorrect to describe section 9(1) as creating a "legal fiction" as a deeming provision does. To the extent that the guidance given by Lord Briggs JSC (with whom the other Justices agreed) is nevertheless helpful by analogy, we note that he emphasised the importance of construing the provision in its context to ensure that it does not produce effects "clearly outside" the purpose for which it is included in the legislation. It should not be applied "so far as to produce unjust, absurd or anomalous results": para 27.

97. The parties also drew the court's attention to the fact that section 9(1) states first that, on the issue of a GRC, a person's *gender* becomes for all purposes the acquired *gender* and then, in parentheses, that the person's *sex* becomes that of the acquired *sex*. We do not draw any inference from this as to the intended breadth of the rule set out in section 9(1). In our judgment, the words in parenthesis are more likely to be

intended to forestall any argument that might have arisen if the rule referred only to gender and not to sex (or only to sex and not to gender) and to reflect the fact that the words "gender" and "sex" were used interchangeably in legislation at the time the GRA 2004 was introduced. As Lord Reed PSC said in *R (Elan-Cane) v Secretary of State for the Home Department* [2021] UKSC 56, [2023] AC 559, legislation across the statute book assumes that all individuals can be categorised as belonging to one of two sexes or genders and those terms have been used interchangeably: para 52.

98. We can address some preliminary points that were raised by the parties.

99. The appellant submitted that the usefulness of section 9(1) was now spent because the problems encountered by trans men and trans women that the legislation was designed to remove have all been removed by other legislation. The pension age for men and women has now been equalised and gender distinctions in many social security benefits have been removed. Civil partnerships and marriage can now be validly entered into by same sex as well as different sex couples. Given the diminished relevance of the GRA 2004 to the rights of transgender people, the appellant argues that the rule in section 9(1) is also largely spent.

100. We do not accept that. Although many provisions of the GRA 2004 have been overtaken by other legislative developments, we consider that the Act continues to have relevance and importance in providing for legal recognition of the rights of transgender people. This recognition of their changed status has practical effects for individual rights and freedoms (including, for example, in the context of marriage, pensions, retirement and social security) but also in recognising their personal autonomy and dignity and avoiding unacceptable discordance in their sense of identity as a transgender person living in an acquired gender. We also agree with the Scottish Ministers that the GRA 2004 is concerned with relationships between private parties as well as between the transgender person and the state.

101. We do, however, see force in Mr Cooper's argument that the carve out in section 9(3) is not limited to express statutory provision excluding the application of section 9(1) or to circumstances where that is a necessary implication. The "necessary implication" test was discussed by the House of Lords in *R (Morgan Grenfell & Co Ltd) v Special Commissioner of Income Tax* [2002] UKHL 21, [2003] 1 AC 563. That case concerned the abrogation of an important common law right, namely the right to rely on legal professional privilege to resist a request for disclosure of documents. At para 45 of his speech, Lord Hobhouse of Woodborough said that where the statute did not contain any express words that abrogated that right, the question arose whether there was a necessary implication to that effect. He described the test to be applied in those circumstances as distinguishing between what it would have been sensible or reasonable for Parliament to have included or what Parliament would, if it had thought about it, probably have included and what it is clear that the express language of the statute shows that the statute must have included: "A necessary implication is a matter of express language and logic not interpretation": para 45.

102. However, the stringency of the necessary implication test is not appropriate when considering the application of section 9(3) of the GRA 2004. The principle of legality described by Lord Hoffmann in *R v Secretary of State for the Home Department, ex p Simms* [2000] 2 AC 115, 131, is not engaged here. This is not a case where the court is being

asked to override a basic tenet of the common law or constitutional rights. We therefore reject the submissions made by Ms Irvine for the Scottish Ministers and by the EHRC that only express wording or necessary implication applying that stringent test can disapply the rule in section 9(1).

103. We also reject the submission that the carve out in section 9(3) only operates in respect of future legislation and not legislation, such as the SDA 1975, which was already enacted at the date when the GRA 2004 was enacted. Ms Irvine submitted that the GRA 2004 itself made exhaustive provision for how the rule was to apply to existing statutes. We do not accept that; section 9(3) refers to "any other enactment" and those words have a clear meaning.

104. It is true, as Mr Coppel appearing for the EHRC pointed out, that the explanatory notes for the GRA 2004 described section 9(3) as meaning that the general proposition in section 9(1) was subject to exceptions made by the Act itself "and, for the future, by any other enactment or subordinate legislation" (para 29). But that is not what section 9(3) says and we conclude that the notes are in error in this regard.

105. Limiting the application of section 9(3) to legislation enacted after the GRA 2004 might in some cases produce results adverse to the trans community. For example, *R (McConnell) v Registrar General for England and Wales* [2020] EWCA Civ 559, [2021] Fam 77 concerned a judicial review claim by a trans man with a GRC who had given birth to a son following fertility treatment. The principal issue before the Court of Appeal was whether section 12 of the GRA 2004 which provides that a change of gender "does not affect the status of the person as the father or mother of a child" only precluded recharacterising someone's status as regards children born before the issue of the GRC or whether it also determined the parent's status if the baby was born after the issue of that certificate. The Court held it was both prospective and retrospective so that he had been correctly referred to as the child's "mother" on the birth certificate.

106. The Court of Appeal referred in *McConnell* at paras 23 onwards to an issue that had been raised at first instance before the President of the Family Division, Sir Andrew McFarlane, as to whether Mr McConnell could lawfully have been provided with the fertility treatment which resulted in the birth of his son. In his judgment at first instance ([2019] EWHC 2384 (Fam), [2020] Fam 45), the President noted that the fertility treatment provided by the clinic to Mr McConnell could only lawfully be provided if it comprised "treatment services" capable of being licensed by the Human Fertilisation and Embryology Act 1990 ("HFEA"). "Treatment services" were defined in section 2 of the HFEA as services provided "for the purpose of assisting women to carry children". It is a criminal offence to undertake the creation of an embryo except in pursuance of a licence. The President noted:

> "155. If at the time that he received treatment services at the clinic [Mr McConnell] had been a woman (which by virtue of the GR certificate he was not) then the placing into his womb of gametes, in the form of permitted sperm, would have been lawful under the terms of the clinic's licence, assuming any other licence conditions had been complied with. It must, however, be at least questionable whether the provision of treatment services to a man is within the range of activities that the HFEA is permitted to authorise by licence."

107. The Government's case before the court in *McConnell* was that a decision that Mr McConnell was not a "woman" for the purposes of the scheme would have grave adverse policy consequences. The treatment might be wholly outside the regulatory scheme with the possible result not only that the treatment was unlawful, but that the donor of the sperm became the legal father of the child. The Government also rejected Mr McConnell's argument that the word "woman" in the HFEA now meant "person"; that would "cause insuperable problems elsewhere in the HFEA": para 157. The President declined to determine the issue and the Court of Appeal also did not consider it necessary or appropriate to comment on the question of whether Mr McConnell's treatment was lawfully provided: para 26.

108. This court also does not express any view on that issue. We note only that the effect of the rule in section 9(1) on the very many statutes referring to men and women, whether enacted before or after the GRA 2004, must be carefully considered in the light of the wording, context and policy of the statute in question. It is likely to be unhelpful for the coherence of the law to impose a stringent test for the application of section 9(3).

(v) Case law on the operation of section 9(1)

109. The implications of the change of gender "for all purposes" have been discussed in a number of cases. In *R (C) v DWP* the applicant challenged the policy of the Department of Work and Pensions to retain on its database information about her former (male) sex including her former titles and names. The Supreme Court rejected the argument that the policy was a breach of section 9(1). Lady Hale (with whom the other Justices agreed) said:

> "23. The problem with this argument is that section 9(1) clearly contemplates a change in the state of affairs: before the issue of the GRC a person was of one gender and after the issue of the GRC that person 'becomes' a person of another gender. The sections which follow section 9 are designed, in their different ways, to cater for the effect of that change. …
>
> 24. There is nothing in section 9 to require that the previous state of affairs be expunged from the records of officialdom. Nor could it eliminate it from the memories of family and friends who knew the person in another life."

110. Those passages were cited by the Employment Appeal Tribunal ("EAT") in *Forstater v CGD Europe* [2022] ICR 1 when upholding a complaint of unlawful discrimination by a consultant whose contract had not been renewed because she had expressed gender critical views, that is to say, views supporting the contention that biological sex is immutable. At the end of a comprehensive and impressive judgment, Choudhury P addressed the question whether the presence of section 9(1) meant that the claimant's views were not worthy of protection under section 10 of the EA 2010 and article 9 of the Convention:

> "99. The effect of a GRC, whilst broad as a matter of law, does not mean that a person who, like the claimant, continues to believe that a trans woman with a GRC is still a man, is necessarily in breach of the GRA by doing so; the GRA does not compel a person to believe something that they do not, any

more than the recognition by the state of civil partnerships can compel some persons of faith to believe that a marriage between anyone other than a man and a woman is acceptable. That is not to say, of course, that the claimant can, as a result of her belief, disregard the GRC; clearly, she cannot do so in circumstances where the acquired gender is legally relevant, eg in a claim of sex discrimination or harassment."

111. The EAT commented that if the claimant gave expression to her beliefs by refusing to refer to a trans person by their preferred pronoun, that could amount to unlawful harassment in some circumstances, although it would not always have that effect: see paras 103 and 104.

(9) Equality Act 2010: overview of the purpose of the legislation

112. The EA 2010 is an important piece of legislation with a wide scope. It regulates and conditions the relationships and interactions between private individuals and both private and public entities over a field of activities that ranges from ensuring fair recruitment, pay, and treatment in the workplace, to the regulation of the professions, the protection of students from unlawful discrimination in schools, colleges, and universities and the prevention of unfair treatment when accessing healthcare, membership clubs, associations and other goods and services.

113. It is both an amending and consolidating statute which was intended, among other things, to "reform and harmonise equality law and restate the greater part of the enactments relating to discrimination and harassment related to certain personal characteristics" (long title). It consolidated and reformed the Equal Pay Act 1970, the SDA 1975, the Race Relations Act 1976, the Disability Discrimination Act 1995 and other (primarily secondary) legislation addressing unlawful discrimination in other specific areas (religion or belief, sexual orientation, and age) to strengthen the law in order to support greater progress on equality.

(10) The structure of the EA 2010

114. The EA 2010 is arranged over 16 parts and 28 schedules. It closely defines the various forms of prohibited conduct regulated by its provisions. It does that first by establishing "key concepts" in Part 2 and then, in subsequent parts, by creating a series of statutory torts, that is acts that are unlawful conduct in the context of certain activities. These broadly relate to services and public functions; premises; work; education; and associations. The unlawful acts have a wide coverage but there are also numerous exemptions to the unlawful acts created by the EA 2010, some of general application and others specific to certain unlawful acts or characteristics. Individuals can enforce rights under the EA 2010 in tribunals and courts. The EHRC has a role in taking strategic and certain enforcement action under the EA 2010.

115. So far as key concepts are concerned, Part 2 explains and defines the forms of discrimination and other conduct that is prohibited by subsequent parts of the Act. Discrimination comprises:

(a) Direct discrimination (defined in section 13)

(b) Combined discrimination because of a combination of two relevant protected characteristics (defined in section 14)

(c) Discrimination arising from disability (defined in section 15)

(d) Gender reassignment discrimination: cases of absence from work (defined in section 16)

(e) Pregnancy and maternity discrimination: non-work cases (defined in section 17)

(f) Pregnancy and maternity: work cases (defined in section 18)

(g) Indirect discrimination (defined in sections 19 and 19A)
Other prohibited conduct comprises harassment (section 26) and victimisation (section 27) though these are not forms of discrimination for the purposes of the EA 2010.

116. Section 25 is headed "References to particular strands of discrimination" and sets out what is meant by references to characteristic specific discrimination. It covers the nine "protected characteristics" under the EA 2010, namely: age, disability, gender reassignment, marriage and civil partnership, pregnancy and maternity, race, religion or belief, sex and sexual orientation. Relevantly for our purposes, it provides that:

"(3) Gender reassignment discrimination is—

(a) discrimination within section 13 because of gender reassignment;

(b) discrimination within section 16;

(c) discrimination within section 19 or 19A where the relevant protected characteristic is gender reassignment.

(4) Marriage and civil partnership discrimination is—

(a) discrimination within section 13 because of marriage and civil partnership;

(b) discrimination within section 19 or 19A where the relevant protected characteristic is marriage and civil partnership.

(5) Pregnancy and maternity discrimination is discrimination within section 17 or 18. ...

(8) Sex discrimination is—

(a) discrimination within section 13 because of sex;

(b) discrimination within section 19 or 19A where the relevant protected characteristic is sex.

(9) Sexual orientation discrimination is—

(a) discrimination within section 13 because of sexual orientation;

(b) discrimination within section 19 or 19A where the relevant protected characteristic is sexual orientation."

117. These key concepts are then applied in the subsequent parts of the EA 2010 which set out what conduct is unlawful and the exemptions that can be invoked in certain circumstances. We summarise them by way of overview in the next 11 paragraphs.

118. Part 3 (see also Schedules 2 and 3) makes it unlawful (in relation to all protected characteristics except marriage and civil partnership – section 28(1)(b) – and age so far as relating to those under 18 – section 28(1)(a)) to discriminate against (in other words, directly or indirectly), or to harass or victimise a person when providing a service (which is defined in section 31 to include the provision of goods or facilities) or when exercising a public function. Section 29 is the central provision in this Part and provides:

"29. (1) A person (a 'service-provider') concerned with the provision of a service to the public or a section of the public (for payment or not) must not discriminate against a person requiring the service by not providing the person with the service.

(2) A service-provider (A) must not, in providing the service, discriminate against a person (B)—

(a) as to the terms on which A provides the service to B;

(b) by terminating the provision of the service to B;

(c) by subjecting B to any other detriment. ..."

(Subsections (3), (4) and (5) prohibit harassment and victimisation by a service-provider.)

119. Part 4 makes the same or similar provision in relation to persons disposing of (for example, by selling or letting) or managing premises.

120. Part 5 (see also Schedules 6, 7, 8 and 9) makes it unlawful to discriminate against, harass or victimise a person at work or in some forms of employment. It regulates prohibited conduct against prospective, existing and former employees and other workers, and applies to all employers, to partnerships, the police, the Bar, advocates, officeholders, appointments (etc) to public offices, qualification bodies, employment service-providers, trade organisations, and local authorities. In the ordinary employment context, section 39(1) and (2) prohibits discrimination by employers against applicants for employment and employees, in summary, in deciding who should be offered employment, in the terms of employment afforded, in dismissing a person or in subjecting that person to any other detriment.

121. This Part also contains provisions relating to equal pay between men and women.

By section 64, these provisions (sections 66 to 70) apply where, "(1) … (a) a person (A) is employed on work that is equal to the work that a comparator of the opposite sex (B) does; (b) a person (A) holding a personal or public office does work that is equal to the work that a comparator of the opposite sex (B) does". Section 65 defines equal work for these purposes, including where work is rated as equivalent by a job evaluation study or where it would have been rated as equal were the evaluation not made on a "sex-specific system" (ie a system that sets values on demands made on a worker that are "for men different from those it sets for women"): see section 65(4) and (5).

122. It also applies to pregnancy and maternity equality "where a woman … is employed, or … holds a personal or public office" (section 72). For example, by section 73(1) if the "terms of the woman's work do not (by whatever means) include a maternity equality clause" they are treated as including one. A maternity equality clause is a provision that, "in relation to the terms of the woman's work", has the effects provided for by section 74. So far as concerns membership rules or rights under an occupational pension scheme, section 75(1) treats such a scheme as including a maternity equality rule if one is not included in the scheme.

123. Provision is also made in this Part making it unlawful for an employment contract to prevent an employee disclosing his or her pay to a colleague; and a power to require private sector employers to publish gender pay gap (the size of the difference between men and women's pay expressed as a percentage) information about differences in pay between men and women (see sections 77 and 78).

124. Part 6 makes it unlawful for education bodies (including higher and further education) to discriminate against, harass or victimise a school pupil or student or applicant for a place.

125. Part 7 (see also Schedules 15 and 16) makes it unlawful for associations (for example, private clubs and associations or other organisations) to discriminate against, harass or victimise members, associates or guests. For the purposes of this Part, associations are defined by section 107(2) as follows: "(2) An 'association' is an association of persons— (a) which has at least 25 members, and (b) admission to membership of which is regulated by the association's rules and involves a process of selection".

126. Provisions for enforcement of the obligations imposed by the EA 2010 are in Part 9. It is unnecessary to elaborate on these for the purposes of this appeal.

127. Part 11 (see also Schedules 18 and 19) establishes a general duty on public authorities to have due regard, when carrying out their functions, to the need: to eliminate unlawful discrimination, harassment or victimisation; to advance equality of opportunity; and to foster good relations: section 149. This section creates what is known as a public sector equality duty or "PSED". It also contains provisions which enable an employer or service-provider or other organisation to take positive action to overcome or minimise a disadvantage arising from people possessing a relevant protected characteristic: sections 158 and 159.

128. Part 14 (see also Schedules 22 and 23) establishes exceptions to certain prohibitions in the earlier Parts in relation to a range of conduct, including action required by an

enactment; protection of women; the provision of benefits by charities; and in sport and sporting competitions.

129. We return to several of these provisions below.

(11) The relevant protection afforded by the EA 2010 to individuals and groups

130. The EA 2010 operates to protect both individuals and groups of people who share a protected characteristic from unlawful discrimination. It has been amended from time to time since its enactment, most recently by the Equality Act 2010 (Amendment) Regulations 2023 (SI 2023/1425) (which inserted section 19A referred to below). The version we refer to below is the version as in force at the date of this judgment.

(i) For individuals

131. At an individual level it does so primarily by means of a general prohibition against less favourable treatment in the form of overt or direct discrimination (section 13) because of one or a combination of individual protected characteristics or by means of more specific direct discrimination provisions which operate similarly in relation to particular characteristics (of relevance here are sections 16 in relation to gender reassignment, and sections 17 and 18 in relation to pregnancy and maternity discrimination).

132. Section 13(1) provides the general definition for direct discrimination as follows:

"(1) A person (A) discriminates against another (B) if, because of a protected characteristic, A treats B less favourably than A treats or would treat others."

133. There are three points to note about section 13 at this stage.

134. First, to demonstrate less favourable treatment in subsection (1) an actual or hypothetical comparator is often relied on to demonstrate that a person without the relevant protected characteristic was or would have been treated more favourably by person A. Such a comparator (actual or hypothetical) must be a person who does not share B's protected characteristic. Section 23(1) makes clear that, apart from the protected characteristic, there must be "no material difference between the circumstances relating to each case" when determining whether B has been treated less favourably. Accordingly, where sex is the protected characteristic, a woman relying on section 13(1) must compare her treatment with the treatment that was or would have been afforded to a man whose circumstances are not materially different to hers; in other words, a similarly situated man. Where gender reassignment is the protected characteristic, in the case of a male person proposing to or undergoing gender reassignment to the opposite sex, the correct comparator is likely to be a man without the protected characteristic of gender reassignment and similarly for a woman (although there may be situations where the comparator's sex is immaterial to the comparison). See for example, *Croft v Royal Mail Group plc* [2003] EWCA Civ 1045, [2003] ICR 1425 at para 74.

135. Secondly, pregnancy and maternity are a special category in the sense that there

is no need for any comparison in treatment to be made in the case of pregnancy and maternity discrimination. For this category direct discrimination is defined by reference to unfavourable (not less favourable) treatment. So, for example, in a non-work context, section 17 provides:

"(2) A person (A) discriminates against a woman if A treats her unfavourably because of a pregnancy of hers.

(3) A person (A) discriminates against a woman if, in the period of 26 weeks beginning with the day on which she gives birth, A treats her unfavourably because she has given birth.

(4) The reference in subsection (3) to treating a woman unfavourably because she has given birth includes, in particular, a reference to treating her unfavourably because she is breast-feeding."

136. Similar provision is made by section 18 in a work context:

"(2) A person (A) discriminates against a woman if, in or after the protected period in relation to a pregnancy of hers, A treats her unfavourably— (a) because of the pregnancy, or (b) because of illness suffered by her in that protected period as a result of the pregnancy.

(3) A person (A) discriminates against a woman if A treats her unfavourably because she is on compulsory maternity leave or on equivalent compulsory maternity leave.

(4) A person (A) discriminates against a woman if A treats her unfavourably because she is exercising or seeking to exercise, or has exercised or sought to exercise, the right to ordinary or additional maternity leave or a right to equivalent maternity leave."

137. These provisions recognise that biological men cannot become pregnant and that no comparison can therefore be made between the case of a sick man and a pregnant woman, both of whom need a period of absence from work. The differential provision made for pregnancy and maternity follows from the jurisprudence of the European Court of Justice in the 1990s establishing that since "only women can be refused employment on grounds of pregnancy" a refusal to employ a pregnant woman "therefore constitutes direct discrimination on grounds of sex" without more: see *Dekker v Stichting Vormingscentrum voor Jong Volwassenen (VJV-Centrum) Plus* (Case C-177/88) [1990] ECR 1-3941 at para 12 and *Handels-og Kontorfunktionærernes Forbund i Danmark v Dansk Arbejdsgiverforening* (Case C-179/88) [1990] ECR 1-3979 which held that the dismissal of a woman because she was pregnant constituted direct discrimination on grounds of her sex without any need to compare her circumstances with those of a man.

138. Consistently with sections 17 and 18 of the EA 2010, special provision is made in relation to direct discrimination in section 13(6) of the EA 2010, where the direct discrimination is because of the protected characteristic of sex, as follows:

"(6) If the protected characteristic is sex –

(a) less favourable treatment of a woman includes less favourable treatment of her because she is breast-feeding;

(b) in a case where B is a man, no account is to be taken of special treatment afforded to a woman in connection with pregnancy, childbirth or maternity."

139. In other words, a woman can complain of direct discrimination based on less favourable treatment of her because she is breast-feeding; and a man cannot complain about the "special treatment" afforded only to women in connection with pregnancy, childbirth or maternity.

140. Thirdly, the language of direct discrimination in section 13(1) is different from the language used in the corresponding provision made by section 1(1)(a) of the SDA 1975 which defined direct sex discrimination as treatment by another person of a woman in relevant circumstances if "(a) on the ground of her sex he treats her less favourably than he treats or would treat a man". Section 13(1) by contrast is framed by reference to less favourable treatment "because of a protected characteristic". Under section 13(1) of the EA 2010 therefore the complainant need not herself possess the protected characteristic relied on: see *Coleman v Attridge Law* (Case C-303/06) [2008] ICR 1128, which held that the EU Equal Treatment Framework Directive (2000/78) protects those who, although not themselves disabled, nevertheless suffer direct discrimination or harassment because of their association with a disabled person. Accordingly, the section 13(1) prohibition includes direct discrimination based on perception, whether or not shared by the person being perceived, and by association. We return to this point below (see paras 250 to 257).

141. Gender reassignment discrimination is covered by section 13 save in cases of absence from work. In these cases, different protection is provided in section 16, which defines direct discrimination in relation to cases of absence from work because of gender reassignment, as follows:

"16 Gender reassignment discrimination: cases of absence from work

(1) This section has effect for the purposes of the application of Part 5 (work) to the protected characteristic of gender reassignment.

(2) A person (A) discriminates against a transsexual person (B) if, in relation to an absence of B's that is because of gender reassignment, A treats B less favourably than A would treat B if—

(a) B's absence was because of sickness or injury, or

(b) B's absence was for some other reason and it is not reasonable for B to be treated less favourably.

(3) A person's absence is because of gender reassignment if it is because the person is proposing to undergo, is undergoing or has undergone the process (or part of the process) mentioned in section 7(1)."

(ii) The protection from discrimination afforded on a group basis

142. The EA 2010 is also concerned to prohibit disguised discrimination which operates at a group level. This is important as Michael Foran explains (in an article entitled "Defining Sex in Law" (2025) 141 LQR 76, 91–92:

> "Arguments concerning the definition of a protected characteristic are never simply manifestations of individual claims. They are always group orientated. The claim that one is a woman is a claim to be included within a particular category of persons and to be excluded from another. It is also a claim to include some persons and to exclude other persons within the group that one is a part of. This matters especially for aspects of the Equality Act 2010 which require duty-bearers to be cognisant of how their conduct might affect those who share a protected characteristic or where there is an obligation to account for the distinct needs and interests of those who share a particular characteristic."

143. The group-based protections are aimed at achieving substantive equality of results for groups with a shared protected characteristic. The EA 2010 does this in several different ways, the most significant of which for our purposes are as follows.

144. First, the provisions concerning indirect discrimination are specifically directed at the problem of group discrimination and their purpose is to counter group (not individual) disadvantage. They operate where an apparently neutral policy or practice is applied generally to everyone but produces a disproportionate disadvantage for a particular group with a shared protected characteristic. Indirect discrimination is defined by sections 19 and/or 19A of the EA 2010. Section 19(1) and (2) provide that indirect discrimination occurs when a person (A) applies to another (B) a "provision, criterion or practice" (generally referred to as a "PCP") if:

> "(a) A applies, or would apply, it to persons with whom B does not share the characteristic,
>
> (b) it puts, or would put, persons with whom B shares the characteristic at a particular disadvantage when compared with persons with whom B does not share it,
>
> (c) it puts, or would put, B at that disadvantage, and
>
> (d) A cannot show it to be a proportionate means of achieving a legitimate aim."

145. For indirect discrimination to be established, it must be possible to reach general conclusions or make general assumptions about a group with a particular protected characteristic such that an employer or other duty-bearer ought reasonably to be able to appreciate that any particular PCP applied to their workforce or service users may have a disproportionately adverse impact on the group. For example, an employer, who sets a minimum height requirement for employment in a police service at a level that (for physiological reasons) most men, but fewer women can comply with, can reasonably expect that women as a group will be disadvantaged by such a requirement. Unless the employer can justify the minimum height requirement, it will be unlawful as indirectly sex discriminatory. The same would be true of an employer's requirement

to fulfil the full range of shift patterns as a tube train driver arranged on a 24 hour per day, seven days per week basis, because it is recognised that those with primary child-care responsibilities (for societal reasons, mostly women) will find it harder to comply with such a requirement than those without (mostly men). In both cases, the employer can reasonably be expected to anticipate these consequences for women as a group in the workforce and can therefore be expected to justify the PCP before imposing it.

146. Section 19A is headed "Indirect discrimination: same disadvantage" and extends the scope of the protection against indirect discrimination (as defined in section 19) to cases where a PCP is applied by A to B; and A also applies the PCP to persons who share a relevant protected characteristic and to those who do not share it; and "B does not share that relevant protected characteristic" (section 19A(1)(c)) but B is put at "substantively the same disadvantage as persons who do share the relevant protected characteristic" (section 19A(1)(e)). We return to this provision below at para 259.

147. Secondly, there are provisions which allow for positive action to address disadvantages faced by groups of people with a shared protected characteristic compared to those without that protected characteristic. Thus, employers or organisations can implement measures to improve opportunities for underrepresented groups, provided these measures are proportionate: see for example, sections 158 and 159 of the EA 2010. In broad terms, these provisions enable a person (P) who reasonably thinks that persons who share a protected characteristic either suffer a disadvantage connected to the characteristic, or have needs that are different from the needs of persons who do not share it, or their participation in an activity is disproportionately low, to take proportionate positive action with the aim of enabling or encouraging that group (comprising persons who share the protected characteristic) to overcome or minimise the disadvantage, meet those needs or participate in the activity.

148. The PSED in section 149 of the EA 2010 also requires public authorities or other persons exercising public functions to have due regard to the need to advance equality of opportunity between people with and without protected characteristics, taking account of any particular disadvantages, needs or low participation levels, and to foster good relations between such groups. These provisions are directed at increasing equality of opportunity (see the long title to the EA 2010) and are concerned with the same sort of group disadvantage as the provisions dealing with indirect discrimination, whether referable to differences between different groups of people, or to societal attitudes or structures.

149. Thirdly, the EA 2010 also enforces the principle of equal pay for equal work, requiring employers to ensure that men and women are paid the same for doing equivalent roles or work of equal value. Section 64(1) EA 2010 provides that:

"(1) Sections 66 to 70 apply where – a person (A) is employed on work that is equal to the work that a comparator of the opposite sex (B) does; …"

150. Hypothetical comparators are not permitted in an equal pay claim; A must identify B, a person of the opposite sex, as an actual comparator.

(12) The importance of clarity and consistency both for those with legal rights and protections, and those who have duties imposed on them by the EA 2010

151. Accordingly, it is clear from the above that the EA 2010 gives important legal rights to individuals and groups who are vulnerable to unlawful discrimination because of a particular or shared protected characteristic, and both protects against unlawful discrimination and seeks to advance equal treatment. In doing so it seeks to strike a balance between the rights of one group and another, rights that can conflict with or contradict one another in some circumstances. An obvious example of such conflict emerges in employment cases concerning the protected characteristics of religion or belief on the one hand and sexual orientation on the other: see for example *Islington London Borough Council v Ladele* [2009] EWCA Civ 1357; [2010] 1 WLR 955 which concerned disciplinary proceedings taken against a designated civil partnership registrar who refused to conduct same sex civil partnership ceremonies in accordance with the Civil Partnership Act 2004 on the ground that such unions were contrary to her orthodox Christian belief that "marriage is the union of one man and one woman for life" (para 7).

152. The EA 2010 also imposes duties on individuals and organisations not to discriminate unlawfully. It does so by regulating the practical day-to-day conduct of public and private sector employers (small, medium and large), service-providers and others in relation to employees, workers, service users and members of the public who have one or more protected characteristics. Since sex as a protected characteristic is a ground for these legal rights, it must be possible for sex to be interpreted in a way that is predictable, workable and capable of being consistently understood and applied in practice by this wide range of duty-bearers.

153. The group-based rights or protections in the EA 2010 recognise that people who share a particular protected characteristic (known or perceived) often have common experiences or needs, whether arising from differences of biology or physiology, or societal expectations or structures affecting their group. These shared experiences or needs can and do give rise to particular disadvantage if they are not met, and they differentiate that group from other groups without the protected characteristic. As we have said, the duties imposed by the EA 2010 require an ability to anticipate that particular rules, policies or practices might affect those who share a protected characteristic and have distinct needs or interests in consequence. Those upon whom the EA 2010 imposes duties (the duty-bearers) must regulate their conduct and practices to avoid unlawful indirect discrimination. Organisations considering taking appropriate positive action measures must be able to identify membership of a disadvantaged group sharing a particular characteristic. Public authorities subject to the duty in section 149 (the PSED) must be able to identify differently affected groups if they are to be able to analyse the features which may disadvantage some groups over others or affect relations between them, in order to analyse the impact of their policies.

154. In short, clarity and consistency about how to identify the relevant groups that share protected characteristics are essential to the practical operation of the EA 2010.

(13) The central question: does the EA 2010 make provision within the meaning of section 9(3) of the GRA 2004 to displace the application of section 9(1)?

155. Against that background, we turn to address the central question in this appeal.

156. To recap, section 9(1) of the GRA 2004, read with section 9(2) and (3), has the effect that the gender of a person with a GRC becomes the acquired gender "for all purposes" so that "if the acquired gender is the male gender, the person's sex becomes that of a man and, if it is the female gender, the person's sex becomes that of a woman", unless there is a specific exception in the GRA 2004 itself or unless the terms and context of an enactment, including a subsequent enactment, demonstrate that there is "provision made" by that enactment pursuant to section 9(3) that negates the effect of section 9(1). In other words, section 9(1) applies unless section 9(3) applies. Section 9(3) will obviously apply where the GRA 2004 or subsequent enactment says so expressly. But express disapplication of section 9(1) is not necessary as we have explained. Section 9(3) will also apply where the terms, context and purpose of the relevant enactment show that it does, because of a clear incompatibility or because its provisions are rendered incoherent or unworkable by the application of the rule in section 9(1).

157. There is no doubt that the EA 2010 was enacted in the knowledge of the existence of the GRA 2004, its known consequences and the case-law which both prompted it (*Goodwin*) and confirmed the GRA 2004 as having remedied the Convention breach (*Grant*). Indeed, the EA 2010 contains an exemption for gender reassignment discrimination in the context of solemnisation of marriage, which refers expressly to the effects of section 9(1) as "[the person's] gender has become the acquired gender under the Gender Recognition Act 2004" (see paragraph 24 of Part 6 and paragraph 25 of Part 6ZA of Schedule 3 to the EA 2010). So, the strongly worded rule in section 9(1) of the GRA 2004 must be taken to apply to the EA 2010 by virtue of section 9(2) unless there is "provision made" in the EA 2010, that disapplies or negates the effect of section 9(1) on the meaning of sex in the EA 2010. If section 9(3) does not apply, then the section 9(1) rule does apply and sex in the EA 2010 must have an extended meaning that includes "certificated sex". If that is the position, then the Scottish Ministers' guidance about the application of the 2018 Act is correct and lawful in making clear that trans women with a GRC can count towards the attainment of the goal of achieving 50% representation of women on the public boards covered by the 2018 Act.

158. There is no provision in the EA 2010 that expressly addresses the effect (if any) which section 9(1) of the GRA 2004 has on the definition of "sex" or the words "woman" or "man" (and cognate expressions) used in the EA 2010. The terms "biological sex" and "certificated sex" do not appear anywhere in the Act. However, the mere fact that the word "biological" is absent from the EA 2010 definition of "sex" is not by itself indicative of Parliament's intention that a "certificated sex" meaning is intended. The same is true of the absence of the word "certificated" in the definition of "sex".

159. In the Outer House, Lady Haldane concluded (at para 53) that section 9(2) of the Victims and Witnesses (Scotland) Act 2014 (as amended by the Forensic Medical Services (Victims of Sexual Offences) (Scotland) Act 2021) can only properly (or fairly) be read to mean biological sex when it uses the term "sex". This is because the purpose of the amendment (introduced by the 2021 Act) was to ensure that section 9(2) of the 2014 Act read as follows: "Before a medical examination of the person is carried out by a registered medical practitioner, the person must be given an opportunity to request that any such medical examination be carried out by a registered medical practitioner of a sex specified by the person". We agree with her analysis: sex as used in this provision must mean biological sex notwithstanding that there is no reference to

biological sex in this provision. The clear statutory intention is to respect the right of a female or male victim of a sexual crime to request same sex care should she or he so wish because it has always been, and still is, well recognised that reasonable objection can be taken to an intimate medical examination by a member of the opposite biological sex. References to sex could only be references to biological sex in context.

160. If the EA 2010 can only be read coherently to mean biological sex, the same result must follow. The question that must therefore be answered is whether there are provisions in the EA 2010 that indicate that the biological meaning of sex is plainly intended and/or that a "certificated sex" meaning renders these provisions incoherent or as giving rise to absurdity. An interpretation that produces unworkable, impractical, anomalous or illogical results is unlikely to have been intended by the legislature.

161. What is necessary therefore is a close analysis of the EA 2010 to identify whether there are indicators within it that demonstrate that section 9(3) of the GRA 2004 applies and displaces the rule in section 9(1). We start by considering the core provisions in the EA 2010 that depend on or relate to "sex" to consider whether as a matter of ordinary language these provisions can only properly be interpreted as meaning biological sex, or whether they are to be interpreted as also extending to include persons living in the opposite acquired gender who have been issued with a GRC (see paras 166 to 209 below). We will then consider the practicability and workability of the duties imposed and protections afforded by the EA 2010 if a "certificated sex" interpretation is adopted (see paras 210 to 246 below). Finally, we will consider whether a "biological sex" interpretation is contra-indicated because it would remove important protection under the EA 2010 from trans people with a GRC (see paras 248 to 264 below).

(14) The meaning of sex in the SDA before and after the enactment of the GRA 2004

162. Before the enactment of the GRA 2004 there is no doubt that references to sex, man and woman in the SDA 1975 were references to biological sex. (See paras 36–53 above.)

163. On enactment of the GRA 2004 as we have explained above, limited changes were made to the SDA 1975, but this was not done by amending the definition of sex and introducing a differently constructed concept of sex. Rather, sex remained biological sex as we have explained in paras 80–82 above.

164. There is no reason to suppose that Parliament intended, by the EA 2010, to introduce a change of substance from the SDA 1975, by introducing a modification to the meaning of sex in accordance with section 9(1) of the GRA 2004. The parties did not draw our attention to anything in the documents which led up to the enactment of the EA 2010 which identified a perceived mischief that needed a change of substance in the law in this regard.

165. But even if we are wrong about that, and we acknowledge that it is difficult to understand the purpose of some of the amendments to the GOQ provisions in the SDA 1975 made by the GRA 2004, the EA 2010 both consolidated and reformed anti-discrimination law. As a self-contained reforming statute, it should be interpreted, if reasonably possible, without recourse to the predecessor provisions. The GOQ

provisions in the SDA 1975 were repealed and many of the exemptions were fundamentally reframed. Thus, whatever the position in relation to the SDA 1975, the focus of our analysis is necessarily on the EA 2010.

(15) Analysis of core provisions of the EA 2010

166. We have set out the approach to statutory construction at paras 8–14 above. Our task is to ascertain the meaning of the words "sex", "woman" and "man" used in the EA 2010, read in their particular context and in light of the wider context and purpose of the anti-discrimination provisions in the EA 2010.

167. The two specific characteristics at the heart of this appeal are "sex" and "gender reassignment". These are maintained as distinct and separately protected characteristics in sections 11 and 7 of the EA 2010 respectively, just as they were in the SDA 1975, as amended.

168. Section 11 of the EA 2010 provides:

"In relation to the protected characteristic of sex—

(a) a reference to a person who has a particular protected characteristic is a reference to a man or to a woman;

(b) a reference to persons who share a protected characteristic is a reference to persons of the same sex."

169. The only other guidance as to the meaning of these expressions is given in the general interpretation provisions in section 212(1) which provide:

"In this Act … 'man' means a male of any age; … 'woman' means a female of any age."

170. In other words, what is made unlawful is sex discrimination against women and men; and the provision in section 212(1) ensures that boys and girls are protected against discrimination connected to their sex.

171. The definition of sex in the EA 2010 makes clear that the concept of sex is binary, a person is either a woman or a man. Persons who share that protected characteristic for the purposes of the group-based rights and protections are persons of the same sex and provisions that refer to protection for women necessarily exclude men. Although the word "biological" does not appear in this definition, the ordinary meaning of those plain and unambiguous words corresponds with the biological characteristics that make an individual a man or a woman. These are assumed to be self-explanatory and to require no further explanation. Men and women are on the face of the definition only differentiated as a grouping by the biology they share with their group.

172. A certificated sex interpretation would cut across the definition of the protected characteristic of sex in an incoherent way. References to a "woman" and "women" as a group sharing the protected characteristic of sex would include all females of any age (irrespective of any other protected characteristic) and those trans women (biological

men) who have the protected characteristic of gender reassignment and a GRC (and who are therefore female as a matter of law). The same references would necessarily exclude men of any age, but they would also exclude some (biological) women living in the male gender with a GRC (trans men who are legally male). The converse position would apply to references to "man" and "men" as a group sharing the same protected characteristic. We can identify no good reason why the legislature should have intended that sex-based rights and protections under the EA 2010 should apply to these complex, heterogenous groupings, rather than to the distinct group of (biological) women and girls (or men and boys) with their shared biology leading to shared disadvantage and discrimination faced by them as a distinct group.

173. Moreover, it makes no sense for conduct under the EA 2010 in relation to sex-based rights and protections to be regulated on a practical day-to-day basis by reference to categories that can only be ascertained by knowledge of who possesses a (confidential) certificate. Some of the practical consequences of a certificated sex definition are described in the case presented by Sex Matters. They state that uncertainty and ambiguity about the circumstances in which it is legitimate to treat (biological) women and girls as a distinct group whose interests need to be considered and protected, have the effect that many organisations now feel inhibited in doing so.

174. The definitions in sections 11 and 212(1) are similar (though not identical) to those in the SDA 1975. Mr O'Neill relied on the contrast between the definition in section 212(1) of the EA 2010 which says, "In this Act ... 'woman' means a female of any age" and the previous provisions in sections 5(2) and 82 of the SDA 1975 which said "'woman' includes a female of any age". We do not consider this change to be significant in context. In both cases, the meaning conveyed is simply to make clear that boys and girls are also included within the definition of man and woman respectively. We do not see the words "*includes*" and "*means*" as sufficiently distinctive to lead to any conclusions about whether the EA 2010 was intended to alter or maintain the position under the SDA 1975 that the terms refer to biological sex only.

175. It is significant, however, that there is only one definition of sex. The concept of sex is of foundational importance in the EA 2010. The words sex and woman appear across different parts of the Act and in many sections. It would be surprising if the words sex and woman were intended to have different meanings in different sections or parts of the EA 2010, as the Inner House concluded, especially given the definitions of "man" and "woman" in section 212(1) of the EA 2010. Indeed, it would offend against the principle of legal certainty and the need for a meaning which is constant and predictable, especially in the context of an Act with the purposes we have identified, and which has such practical everyday consequences for so many individuals and organisations in society.

176. The general rule, as we have said, is that words or terms used more than once in the same legislation are taken to have the same meaning whenever they appear, and the general purpose of an interpretation provision is to fix the meaning of such a word or term throughout the legislation in question. This presumption can be rebutted where the context requires, even where a saving for context does not appear in the definition section. But this is likely to be rare and giving a variable meaning to a defined term is generally only done where it is clear that there is a genuine drafting error resulting in differential usage of the word or term in the text of the legislation: see for example the

observations to this effect in a human rights context in *Secretary of State for Work and Pensions v M* [2004] EWCA Civ 1343, [2006] QB 380 at para 84.

177. As we shall demonstrate, a strong indicator that the words "sex", "man" and "woman" in the EA 2010 have their biological meaning (and not a certificated sex meaning) is provided by sections 13(6), 17 and 18 (which relate to sex, pregnancy and maternity discrimination) and the related provisions. The protection afforded by these provisions is predicated on the fact of pregnancy or the fact of having given birth to a child and the taking of leave in consequence. Since as a matter of biology, only biological women can become pregnant, the protection is necessarily restricted to biological women.

178. The repeated references in these sections, to a woman who has become pregnant or who is breast-feeding only make sense if sex has its biological meaning. These plain, unambiguous words can only be interpreted coherently as references to biological sex, biological females and biological males. Put another way, if the acquisition of a certificate pursuant to section 9(1) of the GRA 2004 applies to these words, so that biological women living as trans men (with a GRC in the male gender) are male, they would nonetheless be excluded from protection when pregnant notwithstanding a continued capacity to become pregnant, and duty-bearers would not be able to claim relevant exemptions in relation to their treatment. The protections include the following:

(a) Section 13(6)(a) (in the context of direct discrimination) provides that "If the protected characteristic is sex" "less favourable treatment of a *woman* includes less favourable treatment of *her* because *she is breast-feeding*". This provision obviously applies to biological women only. As a matter of ordinary language, it necessarily excludes all biological men as a homogenous group or class (whether or not they have a GRC).

(b) Section 17(2) makes provision for circumstances where A "discriminates against a *woman* if A treats *her* unfavourably because of a *pregnancy of hers*". Again, all biological women but only biological women are protected; all biological men are excluded from protection as a matter of ordinary language (whether or not they have a GRC).

(c) Section 17(3) gives protection only to a "woman" "because she has given birth". Biological women only are protected (whether or not they have a GRC). Biological men are excluded.

(d) Section 17(4) gives protection only to a "woman" "because she is breast-feeding". Biological women only are protected. Biological men are excluded (whether or not they have a GRC).

(e) Section 18 gives protections only to a "woman" "because of illness suffered by *her* ... as a result of the *pregnancy*"; "because *she* is on [various forms of maternity leave]" or "because *she* is exercising or seeking to exercise, or has exercised or sought to exercise, the right to [various forms of maternity leave]": section 18(2)(a)-(b), (3), (4). Biological women only are protected. All biological men are excluded (whether or not they have a GRC).

179. There are other provisions in the EA 2010 that have similar effect. For example, sections 73 to 76 provide for further protections for maternity-related pay (sections 73 and 74); and for the operation of a maternity equality rule in relation to a term of an occupational pension scheme that "does not treat time when the woman is on maternity leave as it treats time when she is not", by modifying the term "so as to treat time when she is on maternity leave as time when she is not" (section 75(3)). These provisions expressly apply to women only: see section 72 which provides that they apply only where a woman is employed or holds a personal or public office. Schedule 7 paragraph 2 is similar in providing that a "sex equality clause does not have effect in relation to terms of work affording special treatment to women in connection with pregnancy or childbirth".

180. Schedule 9 is also relevant. It sets out a series of exemptions from the general prohibition in section 39 (prohibiting discrimination by an employer against prospective and existing employees, among other things, in the terms of employment offered or in affording access to opportunities for promotion or other benefits and facilities) which render it lawful for employers to treat employees (and others) with certain relevant protected characteristics in a differential way. Paragraph 17 is an exception relating to the provision of "non-contractual payments to women on maternity leave". It provides:

"17(1) A person does not contravene section 39(1)(b) or (2), so far as relating to pregnancy and maternity, by depriving a woman who is on maternity leave of any benefit from the terms of her employment relating to pay."

181. The points we have made above apply with equal force to this provision. Moreover, that the language of "woman" and "her" is deliberate is underscored by a comparison with the immediately following paragraph, paragraph 18, relating to exceptions for benefits dependent on marital status. Paragraph 18 allows access to benefits dependent on marital status to be restricted or refused on grounds of sexual orientation if they accrued or were payable before 5 December 2005 (the day on which section 1 of the Civil Partnership Act 2004 came into force). It does so by identifying those not covered by the exemption in paragraph (1A) as a "person" who is:

"(1A) ... (a) a man who is married to a woman, or (b) a woman who is married to a man, or (c) married to a person of the same sex in a relevant gender change case.

(1B) The reference in sub-paragraph (1A)(c) to a relevant gender change case is a reference to a case where— (a) the married couple were of the opposite sex at the time of their marriage, and (b) a full gender recognition certificate has been issued to one of the couple under the Gender Recognition Act 2004."

182. Three points follow. The references to "man" and "woman" in paragraph 18(1A)(a) and (b) can only be interpreted by their biological meaning for obvious reasons. The reference to a person in (c) is otherwise rendered meaningless, and the provision is unworkable. Secondly, the draftsperson was fully cognisant of the GRA 2004 and plainly understood its unequivocal consequences in the case of marriage where a GRC has been issued. Where the legislation extends to a person to whom a full GRC has been issued, this is done by express provision referring to a "relevant gender change case" and

not by the implicit adoption of a legally constructed concept of sex as certificated sex. Thirdly, the Scottish Ministers and the EHRC submit that this provision would be unnecessary if "sex" in the EA 2010 meant biological sex only, as the couple concerned would not in fact be married to a "person of the same sex in a relevant gender change case"; they would still be married to a person of the opposite sex. We do not accept this contention. The GRA 2004 unequivocally applies to the law relating to marriage. This is a consequence of the GRA 2004 itself and not a consequence of interpreting sex in the EA 2010 in any particular way.

183. Schedule 9 paragraph 20 relates to insurance contracts. It provides:

> "20 (1) It is not a contravention of this Part of this Act, so far as relating to relevant discrimination, to do anything in relation to an annuity, life insurance policy, accident insurance policy or similar matter involving the assessment of risk if— (a) that thing is done by reference to actuarial or other data from a source on which it is reasonable to rely, and (b) it is reasonable to do it."

184. Relevant discrimination is defined by paragraph 20(2) as including both "pregnancy and maternity discrimination" and "sex discrimination". In the case of pregnancy and maternity discrimination, only women who can become pregnant can be affected by differential provision of this kind, but a certificated sex definition would exclude trans men who are pregnant (that is, pregnant women living as trans men with a GRC). In the case of sex discrimination, it is impossible to see how an assessment of the differential risk known to be posed by, say, women and men drivers, could possibly be made by reference to actuarial or other reliable data sources that had also to take account of certificated sex based on a GRC. There is no rational basis for thinking that having a certificate could make a difference to the risk posed by drivers of different sexes. Here too, sex can only mean biological sex.

185. There are also provisions in the EA 2010 that allow for differential treatment afforded by service-providers and others to protect the health and safety of women generally and pregnant women in particular. In other words, what would otherwise amount to unlawful discrimination in regulated activities is not unlawful by virtue of these provisions. Two examples are sufficient for our purposes:

> (a) The first example concerns pregnant women and consistently with the provisions we have referred to above, provides that a service-provider covered by section 29 of the EA 2010 can refuse to provide a service (or can impose conditions on the provision of the service) to a "pregnant woman" if they reasonably believe that to do so would create a risk to her "health and safety". Schedule 3 paragraph 14(1) and (2) relevantly provides:

> "(1) A service-provider (A) who refuses to provide the service to a pregnant woman does not discriminate against her in contravention of section 29 because she is pregnant if— a) A reasonably believes that providing her with the service would, because she is pregnant, create a risk to her health or safety, (b) A refuses to provide the service to persons with other physical conditions, and (c) the reason for that refusal is that A reasonably believes that providing the service to such persons would create a risk to their health or safety.

(2) A service-provider (A) who provides, or offers to provide, the service to a pregnant woman on conditions does not discriminate against her in contravention of section 29 because she is pregnant if - (a) the conditions are intended to remove or reduce a risk to her health or safety, (b) A reasonably believes that the provision of the service without the conditions would create a risk to her health or safety, (c) A imposes conditions on the provision of the service to persons with other physical conditions, and (d) the reason for the imposition of those conditions is that A reasonably believes that the provision of the service to such persons without those conditions would create a risk to their health or safety."

(b) This exemption is meaningful and workable only if sex has its biological meaning for the reasons given in para 178 above. Since only biological women can become pregnant, the protection for service-providers is limited to "pregnant women". A service-provider can discriminate lawfully against a pregnant woman by refusing her service or imposing conditions on the service provision to protect the health and safety of the pregnant woman. However, since a trans man with a GRC (who retains the capacity to become pregnant) would be legally male on the Scottish Ministers' case, the service-provider would be unable to rely on this provision in Schedule 3 paragraph 14 in order to refuse or place conditions on the provision of services to a person who is a "pregnant man". (The same point is true in relation to the exemption for differential treatment of pregnant women by associations in Schedule 16 paragraph 2 which allows for what would otherwise be unlawful discrimination in contravention of section 101(1)(b) by associations against pregnant women on the grounds of health and safety.)

(c) The second example is of general application to "women" as a group. It concerns Schedule 22 paragraph 2 which is headed "Protection of women". It relates to work and vocational training and provides:

"2(1) A person (P) does not contravene a specified provision only by doing in relation to a woman (W) anything P is required to do to comply with— [a series of enactments concerned with 'the protection of women or a description of women which includes W']. ...

(2) The references to the protection of women are references to protecting women in relation to — (a) pregnancy or maternity, or (b) any other circumstances giving rise to risks specifically affecting women. ..."

186. We have dealt with considerations affecting pregnancy above and they apply equally to paragraph 2(2)(a). Paragraph 2(2)(b) is broader. A certificated sex interpretation would make paragraph 2(2)(b) unworkable: it would be impossible to identify "risks specifically affecting women" because the same health or safety risks would also naturally and inevitably be risks that affect trans men with a GRC who would be legally male on this interpretation (albeit biologically female) and therefore liable to be affected by the same risks.

187. A similar point can be made in relation to the provision made by Schedule 7 paragraph 1 which provides that "Neither a sex equality clause nor a maternity equality

clause has effect in relation to terms of work affected by compliance with laws regulating— (a) the employment of women; (b) the appointment of women to personal or public offices"; and to the provision made by Schedule 9 paragraph 20 in relation to insurance contracts and sex discrimination (see paras 183 and 184 above). In both cases, the need to identify laws or the assessment of risk affecting women as a group is rendered difficult if not impossible by a certificated sex interpretation of sex in these provisions.

188. These provisions and the protection against pregnancy and associated (maternity and breast-feeding) discrimination in the EA 2010 are expressly tied to the plain and unambiguous words "woman", "maternity" and the pronouns "she" and "hers". There are no references to risks specifically affecting men, or to a man (or person) who has become pregnant, requires paternity leave or is breast-feeding. The only reference to a man in this context is in section 13(6)(b) which prevents men from complaining about the special treatment accorded to women in connection with pregnancy or childbirth. These provisions are all incoherent and unworkable unless woman and man have their biological meaning.

(16) No variable definition of woman

189. The Second Division of the Inner House recognised the force of this manifestly obvious conclusion in relation to provisions related to pregnancy and maternity. The Inner House concluded that since pregnancy is a matter of fact which hinges entirely on biology, these provisions do mandate a biological meaning of sex (paras 61 and 62 of the judgment). In reaching that conclusion, the Inner House also recognised that there might be other provisions in the EA 2010 where it might equally be necessary to adopt what they described as a "contextual interpretation" of sex as "based on biology" (para 53). However, the Inner House regarded it as impractical to examine every section and every schedule of the EA 2010 to address this possibility.

190. At para 62 the Inner House continued:

> "To interpret these provisions as including only those who are pregnant both as a matter of fact and biology, regardless of the terms of any GRC, does not detract from the proposition that the default interpretation of 'woman' or 'female' would, elsewhere in the Act, include such a person. We do not consider that such an approach leads to an interpretation which is other than 'constant and predictable' (*Imperial Tobacco*, Lord Hope, para 14). We do not understand the observations in *Imperial Tobacco* as excluding an interpretation under which a word or phrase has a default meaning within a statute other than where the context clearly mandates otherwise. What is required is that whenever the phrase or word occurs, its meaning within the particular context where it appears is clear and predictable. The approach which we have identified achieves that."

191. We respectfully disagree with this conclusion. By its nature a variable definition is neither clear, constant nor predictable. It is the opposite in fact. It is also contradicted by the single definition of sex that fixes its meaning in the EA 2010.

192. As Lord Nicholls explained in *Spath Holme*, p 397: "Citizens, with the assistance of

their advisers, are intended to be able to understand parliamentary enactments, so that they can regulate their conduct accordingly. They should be able to rely upon what they read in an Act of Parliament". Individuals and organisations required to apply the requirements of the EA 2010 in practice should not have to work out which of the variable definitions apply without assistance from the words of the legislation itself. We address below the other practical difficulties that flow from a certificated sex interpretation.

193. Moreover, the suggestion in para 63 of the Inner House judgment that the implications of the grant of a GRC under the GRA 2004 may not have been adequately considered in the passage of the EA 2010 (which underwent various iterations and amendments during its passage from Bill to complex, multi-faceted statute) is contradicted by the careful, specific references to the GRA 2004 (including as referred to at para 157 above). There is no factual foundation for this conclusion. Rather, as the Inner House expressly observed earlier in the judgment (at para 33), "When the EA was passed in 2010 it must be assumed that Parliament was fully cognisant of the purpose, terms and effect of the GRA".

194. The Scottish Ministers support the conclusion that differential provision has been made for the definition of sex in the EA 2010. They advance two purported justifications (para 52 of their written case). First, that requiring a different meaning to be given to the term "woman" in the pregnancy context is not impermissible as a matter of statutory construction. Secondly, they say that without requiring a different meaning to be given to the term "woman" in the pregnancy provisions, a person such as the claimant in *McConnell* (discussed at paras 105–107 above), namely, a pregnant trans man with a GRC who is for legal purposes a "pregnant man", but a biological woman, would be entitled to protection from discrimination under section 13(1) of the EA 2010 on grounds of gender reassignment – "on the basis that, in so far as the protections afforded to 'women' in respect of birth and maternity fall within the regulated activities, he would, in being treated less favourably by being denied those protections, have been directly discriminated against on that ground".

195. We do not regard either point as justifying a variable definition for sex in the EA 2010. The definition of sex is foundational to the EA 2010. The bare assertion that a variable definition is "not impermissible as a matter of statutory construction" falls far short of providing any compelling basis for concluding that a variable definition was intended in section 212(1) or is required. It is simply not plausible to think that the definition of "sex" as used in the pregnancy and maternity-related provisions is the result of a genuine drafting error. There are no circumstances in which a biological male can become pregnant, and no man can therefore ever be an appropriate comparator in a pregnancy discrimination case. As we have explained, the pregnancy discrimination provisions are deliberately framed on the basis of unfavourable rather than less favourable treatment and tied to biological females for this reason. Given the presumption (which has not been rebutted) that section 212(1) provides a single definition of "woman" for the purposes of the EA 2010, it follows that "woman" wherever used in the EA 2010 must have a single, consistent, stable and predictable meaning.

196. If the Scottish Ministers were right and section 9(1) of the GRA 2004 has effect for the definition of sex throughout the EA 2010, this would suggest a legislative intention

to provide protection only for pregnancies of women who do not have a GRC and to exclude persons living in the male gender (biological women) who have a GRC (and so are male on the Scottish Ministers' case) who may become pregnant (as illustrated by the circumstances of the *McConnell* case). It is difficult to see any good reason for such an approach. The Scottish Ministers' second argument about gender reassignment discrimination seeks to address the oddity of this result but is unsatisfactory as a response: on their case a man denied the protections available to women for pregnancy and maternity is most obviously treated differently on grounds of sex and not gender reassignment, but sex discrimination cannot run. Parliament plainly intended biological women to benefit automatically from these protections and it is unlikely to have been the legislative intention that a pregnant trans man with a GRC (legally male but biologically female) should have to pursue gender reassignment discrimination in order to obtain the benefit of such protection, whereas a pregnant trans man without a GRC (legally and biologically female) is automatically entitled to them as a woman.

197. In any event, this alternative argument fails to engage with the more important consequence of the rejection of a variable definition of sex for the Scottish Ministers' arguments. Just as the pregnancy provisions manifestly require sex, woman and man to be interpreted in accordance with the biological meaning of those words, the same is also true of several other provisions to which we have referred above, and to other core provisions which we address below. Properly understood these further provisions would be unworkable, inconsistent and incoherent if they bore a certificated sex meaning as modified by section 9(1) and (2) of the GRA 2004.

(17) Other core provisions: gender reassignment and sexual orientation

198. Two other core provisions support our conclusion thus far. First, the protected characteristic of gender reassignment is defined distinctly from sex, in section 7 of the EA 2010, and there is no conflation of these separate characteristics. Section 7 provides:

"7 Gender reassignment

(1) A person has the protected characteristic of gender reassignment if the person is proposing to undergo, is undergoing or has undergone a process (or part of a process) for the purpose of reassigning the person's sex by changing physiological or other attributes of sex.

(2) A reference to a transsexual person is a reference to a person who has the protected characteristic of gender reassignment.

(3) In relation to the protected characteristic of gender reassignment—

(a) a reference to a person who has a particular protected characteristic is a reference to a transsexual person;

(b) a reference to persons who share a protected characteristic is a reference to transsexual persons."

199. Accordingly, the EA 2010 recognises sex and gender reassignment as distinct and separate bases for discrimination and inequality, giving separate protection to each.

Those who have the protected characteristic of gender reassignment are referred to as "a transsexual person" (section 7(3)(a)), not as a "trans" woman or man. There is no distinction drawn in section 7 or elsewhere between those for whom the relevant process would involve reassignment male to female or female to male. In other words, it is the attribute of proposing to undergo, undergoing or having undergone a process (or part of a process) for the purpose of reassignment, which is the common factor, not the sex into which the person is reassigned.

200. The definition does not depend on having a GRC. There is no reference (as there could have been) to the GRA 2004 or to a GRC. Instead, it is dependent on a process for the "purpose of reassigning the person's sex by changing physiological or other attributes of sex". But the fact that section 7 refers to a process for reassigning sex does not lead to the conclusion that such a process results in a change in the protected characteristic of sex under the EA 2010. Section 7 does not say this; nor is it said elsewhere in the EA 2010. The Scottish Ministers contend that it is inherent in this provision because it contemplates the possibility of a change in the protected characteristic of sex from "man" to "woman" and vice versa for persons who have obtained a full GRC. Again, section 7 does not say so. There is nothing in its wording to suggest that the change referred to is based on obtaining a paper certificate. The critical process on which the section 7 characteristic depends involves a change in physiological or other attributes of what must necessarily be biological sex; but there is nothing to suggest that undergoing such a process changes a person's sex as a matter of law. It does not. Indeed, a full process of medical transition to the opposite gender without obtaining a GRC has no effect on the person's sex as a matter of law.

201. Section 9(1) of the GRA 2004 only applies where a full GRC has been obtained. Nobody suggests that a person with a protected characteristic of gender reassignment is entitled on that basis alone to be treated as if their sex has changed for any legal purposes. Without a GRC a trans woman protected by section 7 of the EA 2010 is male for legal purposes and so too a trans man is female for legal purposes. It is significant therefore, that section 7 is considerably broader in scope and coverage than the category of people with a GRC. Moreover, as we have observed above, the data shows that an overwhelming majority of people (in England, Wales and Scotland) with the protected characteristic of gender reassignment do not have a GRC.

202. Since, as we have explained above, neither possession of a GRC nor the protected characteristic of gender reassignment require any physiological change or even any change in outward appearance, there is no obvious outward means of distinguishing between a person with the protected characteristic of gender reassignment who has a GRC and a person with that characteristic who does not. The only difference between these two groups is possession of a paper certificate and that fact (possessing a GRC) is confidential to the person who has it and subject to stringent restrictions on disclosure (see section 22 of the GRA 2004). The duty-bearer cannot ask whether it has been obtained. There is, accordingly, no way for duty-bearers to distinguish confidently between these two groups when regulating their conduct in accordance with obligations imposed by the EA 2010. Moreover, in either case, the individual's biological sex may continue to be readily perceivable and may form the basis of unlawful discrimination. A person has the protected characteristic of gender reassignment as soon as they propose to undergo the process so it may be that at that stage there is no change in outward appearance.

203. The consequence of an interpretation of sex in the EA 2010 as extending to certificated sex pursuant to section 9(1) and (2) of the GRA 2004 would also create an odd inequality of status between those who share the protected characteristic of gender reassignment but do or do not hold a GRC, with the smaller group (holders of a GRC) given additional rights, and no obvious means of distinguishing between the two groups. We can see no good reason why the legislature should have intended that people with the protected characteristic of gender reassignment should be regarded and treated differently under the EA 2010 depending on whether or not they possess a (confidential) certificate, even though in many (if not most) cases there will be no material distinction in their personal characteristics, either as regards gender identity, or appearance, or as to how they are perceived or treated by others or society at large. The difficulty this interpretation would create for service-providers, employers and other organisations in applying equality law to these groups is obvious. Research referred to by Sex Matters shows that, since it is in practice impossible for organisations to distinguish between people with the protected characteristic of gender reassignment who do and do not have a GRC, many organisations feel pressured into accepting de facto self-identification for the purposes of identifying whom to treat as a woman or girl when seeking to apply the group-based rights and protections of the EA 2010 in relation to the protected characteristic of sex. The result in some cases is that certain women-only groups, organisations, and charities have come under pressure (including from funders and commissioners) to include trans women and policy decisions have been taken simply to accept members or users of the opposite biological sex, either assuming that they hold a confidential GRC or on the basis of self-identification.

204. The second core provision is section 12 of the EA 2010 which defines the protected characteristic of sexual orientation and is framed by reference to orientation towards persons of the same sex, the opposite sex, or either sex. Read fairly, references to sex in this provision can only mean biological sex. People are not sexually oriented towards those in possession of a certificate.

205. Section 12 provides as follows:

"12 Sexual orientation

(1) Sexual orientation means a person's sexual orientation towards— (a) persons of the same sex, (b) persons of the opposite sex, or (c) persons of either sex.

(2) In relation to the protected characteristic of sexual orientation— (a) a reference to a person who has a particular protected characteristic is a reference to a person who is of a particular sexual orientation; (b) a reference to persons who share a protected characteristic is a reference to persons who are of the same sexual orientation."

206. Accordingly, a person with same sex orientation as a lesbian must be a female who is sexually oriented towards (or attracted to) females, and lesbians as a group are females who share the characteristic of being sexually oriented to females. This is coherent and understandable on a biological understanding of sex. On the other hand, if a GRC under section 9(1) of the GRA 2004 were to alter the meaning of sex under the EA 2010, it would mean that a trans woman (a biological male) with a GRC (so legally female) who

remains sexually oriented to other females would become a same sex attracted female, in other words, a lesbian. The concept of sexual orientation towards members of a particular sex in section 12 is rendered meaningless. It would also affect the composition of the groups who share the same sexual orientation (because a trans woman with a GRC and a sexual orientation towards women would fall to be treated as a lesbian) in a similar way as described above in relation to women and girls.

207. Thus, as well as the inevitable loss of autonomy and dignity for lesbians such an approach would carry with it, it would also have practical implications for lesbians across several areas of their lives (as described by Ms Monaghan KC in her written case for the second interveners). Of particular significance is the impact it would have for lesbian clubs and associations governed by Part 7 of the EA 2010, including relatively small associations (they must have at least 25 members and admission must be regulated by the association's rules and involve a process of selection). Part 7 of the EA 2010 prohibits discrimination, harassment and victimisation against applicants for membership, members and their guests, of clubs and associations: sections 101 and 102 of the EA 2010. However, Schedule 16 paragraph 1 allows an association to restrict membership, access to benefits, services and facilities, and access to guests to "persons who share a protected characteristic". In other words, clubs and associations can restrict membership and access to women or to same sex attracted people without contravening sections 101 and 102 of the EA 2010. But there is no exception permitting the exclusion of trans women (biological men) with a GRC (so legally female). Accordingly, if a GRC changes a person's sex for the purposes of the EA 2010, a women-only club or a club reserved for lesbians would have to admit trans women with a GRC (legal females who are biologically male and attracted to women). Evidence referred to by the second interveners suggests that this is having a chilling effect on lesbians who are no longer using lesbian-only spaces because of the presence of trans women (ie biological men who live in the female gender).

208. It is unprincipled to answer this problem by saying, as the Scottish Ministers do, that associations can restrict membership to less than 25 members so that they are not an "association" for the purposes of Part 7. It is also impractical. The Scottish Ministers also suggested in writing that the fact that the members of the association may not be attracted to a particular woman (a trans woman with a GRC who is therefore legally female) or wish to associate with her, does not diminish the protections which they are entitled to in terms of their own protected characteristic of sexual orientation. Even if this is true (which is doubtful) it does not begin to address the chilling effect a certificated sex interpretation appears to have on the ability of lesbians to associate in lesbian-only spaces. The idea that to do so they should seek instead to restrict membership on the basis of "some shared philosophical belief regarding the immutability of sex" (as Ms Crawford KC suggested in argument) demonstrates the incoherence of the Scottish Ministers' position.

209. In short, the core provisions to which we have referred, which refer to sex, man or woman, are not capable of being read fairly and consistently with the terms of section 9(1) and (2) of the GRA 2004 without defeating their purpose and meaning. The definition of these terms contained in section 212(1), when applied in particular to section 11 (the protected characteristic of sex which is at the heart of this case) is not capable of being interpreted on the basis of certificated sex. Rather, sex has its biological meaning throughout this legislation: "woman" always and only means a biological female

of any age in section 212(1). It follows that a biological male of any age cannot fall within this definition; and "woman" does not mean or sometimes mean or include a male of any age who holds a GRC or exclude a female of any age who holds a GRC. To reach any other conclusion would turn the foundational definition of sex on its head and diminish the protection available to individuals and groups against discrimination on the grounds of sex. As we shall explain below, in relation to sex discrimination, an individual will still be entitled to protection against discrimination on the grounds of sex on its biological meaning. Thus, the objective of non-discrimination between the sexes is maintained, while at the same time protecting individuals with a GRC from non-discrimination and without seriously undermining the intention behind the GRA 2004.

(18) Meaning and workability of other provisions

210. We have so far concentrated on the core provisions of sections 7, 11, 12, 13(6) and 17 to 18 of the EA 2010. There are several other provisions that we must address because, contrary to the reasoning and conclusions of the Inner House, they too demonstrate that an interpretation of sex based on certificated sex would render the EA 2010 incoherent and unworkable. In other words, the proper functioning of these provisions depends on a biological interpretation of sex.

(i) Separate and single-sex services

211. Part 3 of the EA 2010 regulates the provision of services and public functions, and we have set out above the terms of the prohibition in section 29 (making it unlawful, among other things, to discriminate in the provision of a service or the exercise of a public function). Schedule 3 contains exemptions from this general prohibition. As we shall explain, some of these permit what would otherwise constitute gender reassignment discrimination but make no similar provision for persons issued with a full GRC. Other provisions permit carve-outs from what would otherwise constitute sex discrimination under the EA 2010. In enacting these exemptions, the intention must have been to allow for the exclusion of those with the protected characteristic of gender reassignment, regardless of the possession of a GRC, in order to maintain the provision of single or separate services for women and men as distinct groups in appropriate circumstances. These provisions are directed at maintaining the availability of separate or single spaces or services for women (or men) as a group – for example changing rooms, homeless hostels, segregated swimming areas (that might be essential for religious reasons or desirable for the protection of a woman's safety, or the autonomy or privacy and dignity of the two sexes) or medical or counselling services provided only to women (or men) – for example cervical cancer screening for women or prostate cancer screening for men, or counselling for women only as victims of rape or domestic violence.

212. So far as sex discrimination is concerned, paragraph 26 of Schedule 3 provides that the provision of separate services for persons of each sex will not constitute unlawful sex discrimination in the provision of services (contrary to section 29) where joint services for both sexes would be less effective and such provision is a proportionate means of achieving a legitimate aim. Paragraph 26 provides:

"(1) A person does not contravene section 29, so far as relating to sex discrimination, by providing separate services for persons of each sex if— (a) a joint service for persons of both sexes would be less effective, and (b) the limited provision is a proportionate means of achieving a legitimate aim.

(2) A person does not contravene section 29, so far as relating to sex discrimination, by providing separate services differently for persons of each sex if— (a) a joint service for persons of both sexes would be less effective, (b) the extent to which the service is required by one sex makes it not reasonably practicable to provide the service otherwise than as a separate service provided differently for each sex, and (c) the limited provision is a proportionate means of achieving a legitimate aim."

213. If sex has its biological meaning in this paragraph, then a service-provider can separate male and female users as obvious and distinct groups. For example, a homeless shelter could have separate hostels for men and women provided this pursued a legitimate aim, which might be the safety and security of women users or their privacy and dignity (and the same for male users). By contrast, if sex means certificated sex, the service- provider would have to allow access to trans women with a GRC (in other words, biological males who are female according to section 9(1)) to the women's hostel. The following practical difficulties would arise. First, it would be difficult or impossible for the service-provider to distinguish between trans women with and without a GRC because, as we have explained, the two groups are often visually or outwardly indistinguishable. Secondly and more fundamentally, it is likely to be difficult (if not impossible) to establish the conditions necessary for separate services for each sex when each group includes persons of both biological sexes. For example, it is difficult to envisage how the condition in paragraph 26(2)(a) (a joint service for persons of both sexes would be less effective) could ever be fulfilled when each sex includes members of the opposite biological sex in possession of a GRC and excludes members of the same biological sex with a GRC. In other words, if as a matter of law, a service-provider is required to provide services previously limited to women also to trans women with a GRC even if they present as biological men, it is difficult to see how they can then justify refusing to provide those services also to biological men and who also look like biological men.

214. Thirdly, it also follows that although the gender reassignment exception in paragraph 28 (see below) does apply, it would be challenging to prove that exclusion of those with the protected characteristic of gender reassignment is a proportionate means of achieving a legitimate aim (for example, protecting the safety of women) when sex means certificated sex and the group includes trans women with a GRC (who are biologically male but legally female) and excludes trans women without a GRC.

215. Paragraph 27 of Schedule 3 ("Single-sex services") presents similar problems if a certificated sex interpretation is adopted. It deals with services provided to one sex only (for example rape or domestic violence counselling, domestic violence refuges, rape crisis centres, female-only hospital wards and changing rooms). It provides:

"(1) A person does not contravene section 29, so far as relating to sex discrimination, by providing a service only to persons of one sex if— (a) any of the conditions in sub-paragraphs (2) to (7) is satisfied, and (b) the limited provision is a proportionate means of achieving a legitimate aim.

(2) The condition is that only persons of that sex have need of the service.

(3) The condition is that— (a) the service is also provided jointly for persons of both sexes, and (b) the service would be insufficiently effective were it only to be provided jointly.

(4) The condition is that— (a) a joint service for persons of both sexes would be less effective, and (b) the extent to which the service is required by persons of each sex makes it not reasonably practicable to provide separate services.

(5) The condition is that the service is provided at a place which is, or is part of— (a) a hospital, or (b) another establishment for persons requiring special care, supervision or attention.

(6) The condition is that— (a) the service is provided for, or is likely to be used by, two or more persons at the same time, and (b) the circumstances are such that a person of one sex might reasonably object to the presence of a person of the opposite sex.

(7) The condition is that— (a) there is likely to be physical contact between a person (A) to whom the service is provided and another person (B), and (b) B might reasonably object if A were not of the same sex as B."

216. The gateway conditions in paragraph 27(2) to (7) cannot be coherently applied if sex does not carry its biological meaning because it is hard to see how the condition in paragraph 27(2) (that only persons of one sex have need of the particular service) can be satisfied if each sex includes members of the opposite biological sex in possession of a GRC and excludes members of the same biological sex with a GRC. For example, take a cervical cancer screening service. On a certificated sex interpretation, a trans man who has a GRC (so is legally male) but (as a biological female) retains a uterus and cervix, has the same need of the cervical cancer screening service that is otherwise reserved for women only. A trans woman with a GRC (so, legally female) would have no such need of that service (as a biological male). The result is that the cervical cancer screening service (needed by biological women only) cannot be said to be needed for members of one sex only on this basis, and condition (2) is not capable of being satisfied.

217. Likewise, a certificated sex interpretation of the conditions in paragraph 27(6) and (7) (that a person of one sex might reasonably object to the presence of a person of the opposite sex, and the physical contact provision) will not be capable of being fulfilled in practice. Again, it is difficult to imagine how or in what circumstances it might be considered reasonable for a woman to object to members of the opposite sex (in condition (6)) where "the opposite sex" would include trans women without a GRC (who remain legally male) but not to "members of her own sex". This would arise if by operation of section 9(1) of the GRA 2004 the group of "members of her own sex" were to include biological men with a GRC, and so legally female who may be physically and outwardly indistinguishable from the former group of trans women without a GRC. While many women in a female-only changing room or on a women-only hospital ward or in a rape counselling group might reasonably object to the presence of biological males, it is difficult to see how the reasonableness of such an objection could be founded on possession or lack of a certificate. This is so especially when the distinction does not

track physical appearance or presentation, and the woman is unlikely to have any information about the GRC at the point at which her objection might be raised. A trans woman with a GRC who presents fully as a woman may feel she is more likely to prompt objections from other users if she enters the men's changing room or other facilities than if she uses the women's changing room or facilities. But in facing that dilemma she is in the same position as a trans woman without a GRC. Although such trans women may in practice choose to use female-only facilities in a way which does not in fact compromise the privacy and dignity of the other women users, the Scottish Ministers do not suggest that a trans woman without a GRC is legally entitled to do so.

218. The physical contact condition (7) gives rise to the same difficulties on a certificated sex interpretation. It can only be met where there is likely to be physical contact between person A, to whom a service is provided, and another person B, and B might reasonably object if A were not of the same sex as B. For example, it is readily understandable that a female massage therapist offering massages in her clients' homes might reasonably object to providing this service to a man in that environment, but for the reasons explained above, hard to see how any reasonable objection to providing the service could depend on whether the trans person (person A) has or does not have a GRC. The objection that B might reasonably have can only fairly be interpreted as being to the biological sex of the other person. It is fanciful (even perverse) to think that any reasonable objection to the presence of a person of the opposite sex could be grounded in GRC status or that a confidential GRC could make any difference at all. Read fairly and in context, the provisions relating to single-sex services can only be interpreted by reference to biological sex.

219. Paragraph 28 provides an additional exception in the context of provision of separate and single-sex services in relation to gender reassignment discrimination. It provides relevantly as follows:

"(1) A person does not contravene section 29, so far as relating to gender reassignment discrimination, only because of anything done in relation to a matter within sub-paragraph (2) if the conduct in question is a proportionate means of achieving a legitimate aim.

(2) The matters are— (a) the provision of separate services for persons of each sex; (b) the provision of separate services differently for persons of each sex; (c) the provision of a service only to persons of one sex."

220. The references in this paragraph only make sense as references to biological sex. Provided it is proportionate, paragraph 28 exempts gender reassignment discrimination but only in the context of the provision of separate services for men and women or single services to one sex. To rely on this exception there must be a separate or single-sex service that satisfies the establishment conditions to which we have just referred (in paragraphs 26 and 27 for example) and as we have observed, these provisions cannot on the face of it operate coherently if provision of services only to persons of one sex means provision of services to a group comprising women (biological females) and trans women with a GRC (biological males but legally female) but not to trans men with a GRC (biological females but legally male). The difficulty of establishing the conditions for a separate or women-only service on an approach tied to certificated sex makes it difficult to envisage any circumstances where the ability to exclude on gender

reassignment grounds could operate.

221. There is nothing in the wording of this provision to indicate that paragraph 28 was directed specifically at those holding a GRC, nor is there any basis for concluding that this is its likely context as the Inner House suggested at para 56. (The example given in the explanatory notes at para 740 also does not distinguish between transsexual people with a GRC and those without: "A group counselling session is provided for female victims of sexual assault. The organisers do not allow transsexual people to attend as they judge that the clients who attend the group session are unlikely to do so if a male-to-female transsexual person was also there. This would be lawful"). We can see nothing to support the Inner House's conclusion that "the importance of this paragraph is that it provides the only basis upon which a person might be permitted to exclude a person with a GRC from services which are provided for their acquired sex". Nor is the EHRC correct to assert that paragraph 28 is redundant on a biological interpretation of sex. On the contrary, if sex means biological sex, then provided it is proportionate, the female only nature of the service would engage paragraph 27 and would permit the exclusion of all males including males living in the female gender regardless of GRC status. Moreover, women living in the male gender could also be excluded under paragraph 28 without this amounting to gender reassignment discrimination. This might be considered proportionate where reasonable objection is taken to their presence, for example, because the gender reassignment process has given them a masculine appearance or attributes to which reasonable objection might be taken in the context of the women-only service being provided. Their exclusion would amount to unlawful gender reassignment discrimination not sex discrimination absent this exception.

(ii) Communal accommodation

222. There is a specific exemption for communal accommodation in Schedule 23, paragraph 3 which allows for both sex discrimination and gender reassignment discrimination as follows:

> "(1) A person does not contravene this Act, so far as relating to sex discrimination or gender reassignment discrimination, only because of anything done in relation to— (a) the admission of persons to communal accommodation; (b) the provision of a benefit, facility or service linked to the accommodation."

223. Communal accommodation is defined as follows:

> "(5) Communal accommodation is residential accommodation which includes dormitories or other shared sleeping accommodation which for reasons of privacy should be used only by persons of the same sex.

> (6) Communal accommodation may include— (a) shared sleeping accommodation for men and for women; (b) ordinary sleeping accommodation; (c) residential accommodation all or part of which should be used only by persons of the same sex because of the nature of the sanitary facilities serving the accommodation."

224. Here too it is plain that sex has its biological meaning. The Inner House however,

held at para 59 that "sex" in this context is defined as including birth sex for those still living in that sex, and "acquired sex" for those in possession of a GRC in the opposite gender. In our judgment, this would undermine the very considerations of privacy and decency between the sexes both in the availability of communal sleeping accommodation and in the use of sanitary facilities which the legislation plainly intended to provide for. If sex has a certificated sex meaning it is difficult to envisage any circumstances in which this gateway could sensibly be met since there would be no rational basis for saying that "for reasons of privacy" any communal accommodation and sanitary facilities should be used by women and trans women with a GRC (so legally female but biologically male) only, but not by trans women without a GRC who may be indistinguishable from those in possession of a GRC (and vice versa). This interpretation would run contrary to the plain intention of these provisions.

225. Accordingly, a certificated sex interpretation produces incoherence in the application of these provisions. Moreover, it is not necessary to achieve the purposes of either the GRA 2004 or the EA 2010. On any view, the plain intention of these provisions is to allow for the provision of separate or single-sex services for women which exclude all (biological) men (or vice-versa). Applying a biological meaning of sex achieves that purpose.

(iii) Single-sex higher education institutions

226. Schedule 12 paragraph 1 addresses admission to single-sex higher education institutions. It provides as follows:

"(1) Section 91(1), so far as relating to sex, does not apply in relation to a single-sex institution.

(2) A single-sex institution is an institution to which section 91 applies, which—(a) admits students of one sex only, or (b) on the basis of the assumption in sub-paragraph (3), would be taken to admit students of one sex only.

(3) That assumption is that students of the opposite sex are to be disregarded if—(a) their admission to the institution is exceptional, or (b) their numbers are comparatively small and their admission is confined to particular courses or classes.

(4) In the case of an institution which is a single-sex institution by virtue of sub-paragraph (3)(b), section 91(2)(a) to (d), so far as relating to sex, does not prohibit confining students of the same sex to particular courses or classes."

(Section 91 prohibits discrimination in relation to the admission and treatment of a student by responsible bodies of education institutions.)

227. Schedule 12 contains no exception for gender reassignment discrimination in respect of single-sex higher education institutions. The Inner House held (para 58) that this did not support the proposition that sex in section 11 of the EA 2010 is to be read as a reference to biological sex. Rather it held that Schedule 12 can simply be read as

circumstances in which Parliament did not consider that an additional carve out for trans people with a GRC was necessary.

228. Again, we respectfully disagree. It was plainly Parliament's intention to allow for single-sex higher education institutions. That much is plain from the express terms of Schedule 12 paragraph 1. However, if sex means certificated sex, the exception from the sex discrimination provisions for single-sex higher education institutions would not allow such institutions to be limited to girls and women, given the absence of any separate exception for gender reassignment discrimination. We can see no rational basis for a certificated sex reading that would oblige such institutions to admit transsexual members of the opposite (biological) sex with a GRC, whose biological sex is likely to be readily identifiable, whilst excluding others without a GRC, whose circumstances may be materially indistinguishable.

(iv) Single characteristic associations and charities

229. Similarly, Schedule 16 paragraph 1 EA 2010 allows for an association to restrict membership to persons who share a protected characteristic (which would otherwise be unlawful discrimination in contravention of section 101(1)(b)). However, if sex means certificated sex, this exception from the sex discrimination provisions for single characteristic associations would not permit such associations with 25 members or more (see section 107(2) of the EA 2010 discussed above) to be limited to biological women. This is because, as we have said, a certificated sex definition of the protected characteristic of sex would include trans women with a GRC.

230. Nor would single-sex charities be able to use the exception in section 193, which allows them to restrict the provision of benefits to persons who share a protected characteristic in pursuance of a charitable instrument. So far as material, section 193 provides:

> "(1) A person does not contravene this Act only by restricting the provision of benefits to persons who share a protected characteristic if— (a) the person acts in pursuance of a charitable instrument, and (b) the provision of the benefits is within subsection (2).

> (2) The provision of benefits is within this subsection if it is— (a) a proportionate means of achieving a legitimate aim, or (b) for the purpose of preventing or compensating for a disadvantage linked to the protected characteristic."

231. Schedule 16 and section 193(1) plainly intend that single-sex associations and charities should be permitted to exist along with other single-characteristic associations. A certificated sex meaning applied to these exceptions would make it impossible for any women's association or charity – including, for example, a mutual support association for women who are victims of male sexual violence, a lesbian social association, a breast-feeding support charity – to be set up or to pursue a dedicated purpose which is directed at the needs of biological females. To require such associations or charities to reconceive of their objects as targeting a group that does not correspond with their original aims, and to allow trans people with a GRC (of the opposite biological sex) to join would significantly undermine the right to associate on the basis of biological sex

(or sexual orientation based on biological sex as we have discussed above).

(v) Women's fair participation in sport etc

232. Section 195 of the EA 2010 is headed "Sport". It provides:

> "(1) A person does not contravene this Act, so far as relating to sex, only by doing anything in relation to the participation of another as a competitor in a gender-affected activity.
>
> (2) A person does not contravene section 29, 33, 34 or 35, so far as relating to gender reassignment, only by doing anything in relation to the participation of a transsexual person as a competitor in a gender-affected activity if it is necessary to do so to secure in relation to the activity— (a) fair competition, or (b) the safety of competitors.
>
> (3) A gender-affected activity is a sport, game or other activity of a competitive nature in circumstances in which the physical strength, stamina or physique of average persons of one sex would put them at a disadvantage compared to average persons of the other sex as competitors in events involving the activity.
>
> (4) In considering whether a sport, game or other activity is gender-affected in relation to children, it is appropriate to take account of the age and stage of development of children who are likely to be competitors."

233. The Scottish Ministers and the EHRC submit that a biological sex reading of this provision makes it partly unnecessary. They contend that on this reading it would only be necessary in relation to single-sex sports to exclude indirect gender reassignment discrimination, rather than both direct and indirect as per section 195(2). We are doubtful that this submission is correct, but in any event, it appears to miss the point. The real question is whether the provision operates coherently or not if a certificated sex interpretation of sex is required to be adopted.

234. We consider that this provision is, again, plainly predicated on biological sex, and may be unworkable if a certificated sex interpretation is required. The exemption it creates is a complete exemption in relation to the prohibition against sex discrimination in sport in relation to the participation of a competitor in a sport that is a gender-affected activity (section 195(1)) and a partial exemption for gender reassignment discrimination in relation to the participation of a transsexual person as a competitor in a gender-affected activity but only where the treatment is necessary for fairness or safety reasons. In both cases the exemption cannot apply unless there is a gender-affected activity. This is a gateway condition.

235. A gender-affected activity is a defined term. It depends on a determination of whether the physical strength, stamina or physique of average persons of one sex would put them at a disadvantage as competitors in a particular sport when compared to average persons of the other sex. Take boxing as an example. This is undoubtedly a gender-affected activity on a biological interpretation of sex in section 195(3). On this basis, it is readily apparent (indeed, obvious) that women's average physical strength,

stamina and/or physique will disadvantage them as competitors against average men in a boxing match. However, if average women as a group for comparison with average men for the purposes of section 195(1) includes trans women with GRCs (so legally female but biologically male) the differences in strength, stamina and physique between the two groups may begin to fade. Although at present the numbers of trans people with GRCs may be statistically insignificant, that could not have been predicted at the time the GRA 2004 was enacted, and the effect of section 9(1) cannot depend on how many people are issued with GRCs. Each group has members of the opposite biological sex in it and the gateway condition may be difficult to establish at all. Even if the gateway condition is established, the approach to the group of trans sportswomen who are potentially to be excluded would differ on a rationally unconnected basis: whether or not they have a paper certificate. To exclude trans women with a GRC from the boxing competition, the organiser would have the additional burden of showing that it was necessary to do so in the interests of fairness or safety, whereas a trans woman without a GRC could simply be excluded as a male under section 195(1).

236. On the other hand, a biological definition of sex would mean that a women's boxing competition organiser could refuse to admit all men, including trans women regardless of their GRC status. This would be covered by the sex discrimination exception in section 195(1). But if, in addition, the providers of the boxing competition were concerned that fair competition or safety necessitates the exclusion of trans men (biological females living in the male gender, irrespective of GRC status) who have taken testosterone to give them more masculine attributes, their exclusion would amount to gender reassignment discrimination, not sex discrimination, but would be permitted by section 195(2). It is here that the gender reassignment exception would be available to ensure that the exclusion is not unlawful, whether as direct or indirect gender reassignment discrimination.

(vi) The public sector equality duty and positive action measures for women

237. We have referred above (at paras 127 and 147-149) to the main terms of the PSED (section 149) and the positive action measures available, both in the workplace (section 159) and in the provision of services (section 158). Other specific provision is made elsewhere – for example, section 104 of the EA 2010 which deals with women only shortlists for Parliamentary seats.

238. As we have explained, all organisations subject to the PSED must have due regard, in considering their rules, policies or practices, to the matters set out in section 149, undertaking where appropriate an equality impact assessment in order to understand how and to what extent the policy in question will affect specific groups with different protected characteristics. Organisations and bodies that are subject to the PSED are required to collect data in order to fulfil this duty.

239. If, in the context of equality between the sexes, the interests of trans women (biological males) who have GRCs (so are legally female) must be considered and advanced as part of the group that share the protected characteristic of being "women", the PSED will require data collection and consideration of a heterogenous group containing biological women, some biological males with a GRC (trans women who are legally female) and excluding some biological females with a GRC (trans men who are legally male). This is a confusing group to envisage because it cuts across and fragments

both biological sex and gender reassignment into heterogenous groupings which may have little in common. Any data collection exercise will be distorted by the heterogenous nature of such a group. Moreover, the distinct discrimination and disadvantage faced by women as a group (or trans people) would simply not be capable of being addressed by the PSED because the group being considered would not be a group that, because of the shared protected characteristic of sex, has experienced discrimination or disadvantage flowing from shared biology, societal norms or prejudice. Whereas the interests of biological women (or men) can be rationally considered and addressed, and likewise, the interests of trans people (who are vulnerable and often disadvantaged for different reasons), we do not understand how the interests of this heterogenous group can begin to be considered and addressed.

240. A similar problem arises in relation to the positive action provisions (addressing particular needs, disadvantages, or under-representation of persons who share a protected characteristic (sections 158-159)). If sex means certificated sex, how can an organisation consider the needs of, or disadvantage to, women separately from men, and if it identifies a need for positive action, must it include trans women with GRCs (but not those without) within that action, and exclude biological females with GRCs?

241. In the case of both sets of provisions, the purpose of addressing the particular needs, disadvantages or participation levels of women as a group with the protected characteristic of sex, is undermined if women as a group includes trans women with a GRC (in other words, biological men who are legally female). The guidance at issue in the present case is a good illustration. If the purpose of the positive action measure is to increase representation on public boards of women (with their shared experience of disadvantage based on sex and overcoming such disadvantage), a certificated sex approach changes the group to be represented. It means that those entitled to be considered for this scheme include biological males who have GRCs but it excludes biological females who have GRCs. This is an irrational approach.

242. Moreover, the different needs of and disadvantages faced by transsexual people (whether or not they have a GRC) can – and in the case of the PSED must – be considered separately without conflating these distinct protected characteristics. To do otherwise is detrimental to both groups. Indeed, a certificated sex reading of sex suggests that the needs and interests of transsexuals without a GRC are different from those with a GRC, though their circumstances may often be indistinguishable. In addressing the need for greater representation of women on public boards, it is hard to see what possible difference it could make to the board in question whether the trans woman in question does or does not hold a GRC.

243. It is no answer to these points to say (as the Scottish Ministers do) that there will always be members of a class who do not conform to the characteristics of the majority of a class and that it does not follow that they are not to be taken as falling within that class and entitled to the benefits to be afforded to it. This misses the point. The group-based rights and duties are concerned with identifying the shared needs and disadvantages that affect women as a group, or trans people as a group. If the first group were to include men and the second group people who are not trans people, it is unlikely that they would have the same needs or share the same disadvantages that would justify their inclusion in the particular group. Equally, the fact that some members of the group do not wish to benefit from a particular measure designed to

reduce, say under-representation of that group, does not mean that they do not share the same needs and disadvantages as the group in question.

244. Accordingly, a certificated sex reading of sex in the EA 2010 is not necessary to meet its purpose in relation to the PSED or positive action provisions. On the other hand, such a reading does both undermine the purposes of those provisions and impede clarity of analysis of the different needs of groups with different protected characteristics under them. These provisions deal with potentially conflicting group interests in the field of equality and discrimination law in which Parliament has chosen to protect sex and gender reassignment as distinct protected characteristics. They do not concern individual rights that affect how transsexuals are treated in their general lives, or their ability to bring claims for any form of unlawful discrimination.

(vii) Armed Forces

245. Schedule 9 paragraph 4 of the EA 2010 provides:

"4(1) A person does not contravene section 39(1)(a) or (c) or (2)(b) by applying in relation to service in the armed forces a relevant requirement if the person shows that the application is a proportionate means of ensuring the combat effectiveness of the armed forces.

(2) A relevant requirement is— (a) a requirement to be a man; (b) a requirement not to be a transsexual person.

(3) This Part of this Act, so far as relating to age or disability, does not apply to service in the armed forces; and section 55, so far as relating to disability, does not apply to work experience in the armed forces."

246. Although we accept that the dual requirement of not being a man and not being transsexual means that this provision can operate effectively on a certificated sex basis, it operates just as effectively adopting a biological sex interpretation. Unlike the Inner House we do not consider that it supports a certificated sex interpretation accordingly.

(19) The EHRC's recognition of problems in their interpretation of sex as certificated sex

247. The EHRC is the expert agency tasked by Parliament with considering the operation of the EA 2010. In its letter of 3 April 2023, the EHRC advised the UK Government about the consequences for the broader functioning of the EA 2010 if the decision of the Lord Ordinary in *For Women Scotland v Scottish Ministers (No 2)* was upheld (as it was in large part, by the Inner House). It is not for the EHRC to determine the meaning of sex in the EA 2010. That is for the courts to do. However, we consider it significant that many of the problems we have identified as leading to incoherence and absurdity in the practical operation of the EA 2010 if a certificated sex interpretation is adopted, are expressly recognised by the EHRC as grounds for urgent consideration of legislative amendment. The letter explains the EHRC's understanding that sex in the EA 2010 has a certificated sex meaning pursuant to sections 9(1) and (2) of the GRA 2004 and continues that "it has not been straightforward for service-providers and employers to apply the law, including in areas such as sport and health services". The EHRC

concludes "that if 'sex' is defined as biological sex for the purposes of EA 2010, this would bring greater legal clarity in eight areas". The eight areas are then discussed as follows (numbering added):

"(1) Pregnancy and maternity: As things stand, protections in the EA 2010 for pregnant women and new mothers fail to cover trans men who are pregnant and whose legal sex is male. Defining 'sex' as biological sex would resolve this issue.

(2) Freedom of association for lesbians and gay men: If sex means legal sex, then sexual orientation changes on acquiring a GRC: some trans women with a GRC become legally lesbian, and some trans men with a GRC become gay men. As things stand a lesbian support group (for instance) may have to admit a trans woman with a GRC attracted to women without a GRC or to trans women who had obtained a GRC. On the biological definition it could restrict membership to biological women.

(3) Freedom of association for women and men: As things stand, a women's book club (for instance) may have to admit a trans woman who had obtained a GRC. On the biological definition it could restrict membership to biological women.

(4) Positive action: Currently, trans women with a GRC could benefit from 'women-only' shortlists and other measures aimed at increasing female participation. Trans men with a GRC could not. A biological definition of sex would correct this perceived anomaly.

(5) Occupational requirements: Employers are sometimes permitted to restrict positions to women or to men. An employer can (for example) require that a warden in a women's or girls' hostel be female. At present, such a role would be open to a trans woman with a GRC, but not to a trans man with a GRC. A biological definition of sex would correct this perceived anomaly.

(6) Single-sex and separate sex services: Service-providers are sometimes permitted to offer services to the sexes separately or to one sex only. For instance, a hospital might run several women only wards. At present, the starting point is that a trans woman with a GRC can access a 'women-only' service. The service-provider would have to conduct a careful balancing exercise to justify excluding all trans women. A biological definition of sex would make it simpler to make a women's-only ward a space for biological women.

(7) Sport: At present, to exclude trans women with a GRC from women's sports, the organiser must show that it was necessary to do so in the interests of fairness or safety. A biological definition of sex would mean that organisers could exclude trans women from women's sport without this additional burden.

(8) Data collection: When data are broken down by legal not biological sex,

the result may seriously distort or impoverish our understanding of social and medical phenomena. A biological definition of sex would require public bodies like universities to apply this category, without the complexity added by a legal definition of sex, to the analysis of data collected in fulfilling the Public Sector Equality Duty."

There are three areas identified by the EHRC where it suggests that a "change" to a biological sex interpretation would be "more ambiguous or potentially disadvantageous". These are as follows, and we discuss each area further below:

"(9) Equal pay provisions: At present, a trans woman with a GRC can bring an equal pay claim by citing a legally male comparator who was paid more. A trans man with a GRC could not. The proposed biological definition would reverse this situation. The effect would be to transfer this right from some trans women to some trans men.

"(10) Direct sex discrimination: At present, a trans woman with a GRC can bring a claim of direct sex discrimination as a woman. A trans man with a GRC could not. The proposed biological definition would reverse this situation. The effect would be to transfer the right from some trans women to some trans men.

"(11) Indirect sex discrimination: At present, a trans woman with a GRC could bring a claim of indirect discrimination as a woman. A trans man with a GRC could not. The proposed biological definition would reverse this situation. The effect would be to transfer this right from some trans women to some trans men....

On balance, we believe that redefining 'sex' in EA 2010 to mean biological sex would create rationalisations, simplifications, clarity and/or reductions in risk for maternity services, providers and users of other services, gay and lesbian associations, sports organisers and employers. It therefore merits further consideration."

(20) Why this interpretation would not be disadvantageous to or remove protection from trans people with or without a GRC

248. Finally, we have concluded that a biological sex interpretation would not have the effect of disadvantaging or removing important protection under the EA 2010 from trans people (whether with or without a GRC). Our reasons for this conclusion follow. We consider protection from both direct and indirect discrimination and harassment, and equal pay.

(i) Direct discrimination and harassment

249. It is now well-established that direct discrimination because of a protected characteristic (section 13 of the EA 2010) encompasses not only cases where the complainant affected by discrimination has the protected characteristic in question, but also where the discriminator perceives that the complainant has the characteristic, or in some other way associates the complainant with the protected characteristic. This can

occur, for example, where the complainant is discriminated against because of caring responsibilities for a person with a protected characteristic, such as disability (as happened in *Coleman v Attridge Law* [2008] ICR 1128; *EBR Attridge LLP v Coleman* [2010] ICR 242) or where the complainant is treated detrimentally because it is thought that she or he has a particular protected characteristic even if they do not (*English v Thomas Sanderson Blinds Ltd* [2009] ICR 543, paras 37 to 40 per Sedley LJ, where the protected characteristic was sexual orientation). What is required is that the protected characteristic is a ground for the treatment in question. Terms such as "associative discrimination" and "discrimination by perception" are not a critical part of the analysis. What matters in the former is whether the treatment of the complainant was done because of the protected characteristic of the other person. In a case of perceived discrimination, the correct comparator is someone who is not perceived to have that protected characteristic: *Chief Constable of Norfolk Constabulary v Coffey* [2020] ICR 145. In *Coffey* the EAT (Judge David Richardson) held that where a claimant is treated less favourably on the basis of a mistaken perception that she was disabled, the correct hypothetical comparator was a person who was not perceived to be disabled and who had the same abilities as the claimant. On appeal, Underhill LJ (who gave a judgment with which the other members agreed) upheld the decision and expressly endorsed this comparator (see para 68 referring to para 66 of the EAT judgment [2018] ICR 812. (The perception-based approach was approved by Lord Mance JSC in *R (E) v Governing Body of JFS* [2009] UKSC 15, [2010] 2 AC 728 at para 85.)

250. Applied in the context of a discrimination claim made by a trans woman (a biological male with or without a GRC), the claimant can claim sex discrimination because she is perceived as a woman and can compare her treatment with that of a person not perceived to be a woman (whether that is a biological male or a trans man perceived to be male). There is no need for her to declare her true biological sex. There is nothing disadvantageous about this approach. Neither a biological woman nor a trans woman "bring a claim of direct sex discrimination *as a woman*" (as the EHRC suggests). That is not how the EA 2010 operates: a person brings a claim alleging sex discrimination because of a protected characteristic of sex.

251. Take, for example, a trans woman who applies for a job as a sales representative and the sales manager thinks that she is a biological woman because of her appearance and does not offer her the job even though she performed best at interview and gives the job instead to a biological man. She would have a claim for direct discrimination because of her perceived sex and her comparator would be someone who is not perceived to be a woman. The fact that she is not a biological woman should make no difference to her claim, which would be treated in the same way as a direct discrimination claim made by a biological woman based on the sex of the complainant herself.

252. The same approach would follow in a claim of discrimination by association: the appropriate comparator is someone not associated with the protected characteristic, so that a trans woman (biologically male) treated less favourably because of her association with women could claim sex discrimination and compare her treatment with someone who was not associated with women in the same way or manner (whether that was a biological male living as a man or a trans man).

253. It follows that a certificated sex reading of sex in the EA 2010 is not necessary to achieve the purposes of either the GRA 2004 or the EA 2010 as regards protection from

direct discrimination. A man who identifies as a woman who is treated less favourably because of the protected characteristic of gender reassignment, will be able to claim on that basis. A man who identifies as a woman who is treated less favourably not because of being trans (the protected characteristic of gender reassignment), but because of being perceived as being a woman, will be able to claim for direct sex discrimination on that basis. This does not entail any practical disadvantage and there is no discordance (as the Scottish Ministers appear to suggest) between the individual's position in society and the ability to claim on this basis. A certificated sex reading of the EA 2010 is not necessary here, and the approach applies equally whether or not the claimant has a GRC.

254. The same is true of harassment pursuant to section 26 of the EA 2010. Harassment is defined by section 26 which provides:

"(1) A person (A) harasses another (B) if—

(a) A engages in unwanted conduct related to a relevant protected characteristic, and

(b) the conduct has the purpose or effect of—

(i) violating B's dignity, or

(ii) creating an intimidating, hostile, degrading, humiliating or offensive environment for B."

255. To establish harassment, it is simply necessary to establish a sufficient link between the unwanted conduct and a relevant protected characteristic, and that the conduct violates dignity or creates an intimidating, hostile, degrading, humiliating or offensive environment for the claimant. It follows that, as with section 13(1) EA 2010, under section 26(1)(a) the complainant, B, does not have to possess the relevant protected characteristic to bring an unlawful harassment claim. Conduct will fall within this section where it is related to B's own protected characteristic, or where it is related to a relevant protected characteristic of another person or persons.

256. Applied, for example, to the case of a trans woman with a GRC, who presents as a woman at work and is perceived as a woman, and whose trans status and GRC are confidential: if a colleague harasses her (by making sexualised references to what she is wearing, or degrading comments about how she looks) she can bring a claim for harassment related to sex. She can also bring a harassment claim related to the protected characteristic of gender reassignment but may not wish to do so.

257. Conversely, if a certificated sex reading were adopted, it would have the effect of removing an important aspect of group protection for men and women in the way that direct discrimination under section 13 has been understood to operate. It is well-established that where a policy or rule is applied which applies a criterion that is indissociable from sex in order to determine entitlement to some benefit, that will necessarily constitute unlawful direct discrimination that cannot be justified. A clear example is the policy adopted by the council in *James v Eastleigh Borough Council* [1990] 2 AC 751 (see p764A per Lord Bridge) regarding free admission to the swimming pool

for those of pension age at a time when pension ages for men and women were different. For this principle to apply, there must be an "exact correspondence" between the protected characteristic and the criterion in question (see *Preddy v Bull* [2013] UKSC 73, [2013] 1 WLR 3741 at para 21 per Lady Hale DPSC). A certificated sex reading of sex in the EA 2010 would have the effect of preventing that principle from applying in relation to a criterion which is indissociable from biological sex (for example, a sex-based biological characteristic) because that criterion would not be indissociable from the more complex grouping that would then constitute members of the relevant "sex" as modified by section 9(1). Instead, the application of such a criterion would fall to be considered as a case of indirect discrimination, with the potential for a justification defence. A certificated sex reading of sex would therefore remove this important aspect of protection in relation to direct discrimination under section 13. It is difficult to see why the GRA 2004 could have intended to remove such protection.

(ii) Indirect discrimination

258. Pursuant to section 19 of the EA 2010, unlawful indirect discrimination occurs where the discriminator applies a PCP which places the claimant and persons who share the same protected characteristic at a particular disadvantage, and the treatment in question cannot be justified. Section 19A extends that protection to persons who do not share the same protected characteristic but suffer the same disadvantage as those who do (section 19A(1)(e)).

259. Section 19A was introduced with effect from 1 January 2024 by the Equality Act 2010 (Amendment) Regulations 2023 (made under sections 12(8) and 13 of the Retained EU Law (Revocation and Reform) Act 2023) in order to preserve the effect of EU law, and in particular, to reproduce the principle established by the Court of Justice of the European Union ("the CJEU") in *CHEZ Razpredelenie Bulgaria AD v Komisia za zashtita ot diskriminatsia* (Case C-83/14) [2015] IRLR 746. In that case, the CJEU held that the principle of discrimination by association extends to both direct and indirect discrimination, so that where a group which shares a protected characteristic is put at a particular disadvantage, a person who is also put at that same disadvantage may claim discrimination even if she does not share the characteristic in question (paras 56–60). (See also *British Airways plc v Rollett* [2024] EAT 131, [2025] ICR 242 where Eady J, President of the EAT, confirmed that, prior to the UK's exit from the EU, the EA 2010 would have been interpreted so as to give effect to *CHEZ* so that a claimant need not have the same protected characteristic as the disadvantaged group to bring an indirect discrimination claim (para 61).)

260. Consequently, transgender people (irrespective of whether they have a GRC) are protected by the indirect discrimination provisions of the EA 2010 without the need for a certificated sex reading of the EA 2010, both in respect of any particular disadvantage suffered by them as a group sharing the characteristic of gender reassignment and, where members of the sex with which they identify are put at a particular disadvantage, insofar as they are also put at that disadvantage. Again, this does not entail any practical disadvantage or involve any discordance between the claim and the individual's position in society. On the contrary, the claim will be founded on the facts of a particular shared disadvantage. Transgender people are also protected from indirect discrimination where they are put at a particular disadvantage which they share with members of their biological sex.

261. Therefore, a certificated sex reading is not required to achieve any relevant purpose of either statute in respect of indirect discrimination. Conversely, if sex means certificated sex, this would undermine the ability to conduct a robust analysis of biological women (or men) as a group with a shared characteristic (see paras 172–173 above). In short, it would entail concluding that Parliament did not intend biological women (or men) to be a distinct protected group within its core indirect discrimination provision.

(iii) Equal pay

262. The EHRC says (in the letter of 3 April) that on a certificated sex interpretation of sex, a trans woman with a GRC can bring an equal pay claim by identifying a male comparator who was paid more than her whereas a trans man with a GRC could not. This is true. But the position would simply reverse if either the trans man or trans woman did not have a GRC: in other words, a trans man with a GRC (legally male but biologically female) cannot rely on a male comparator to bring an equal pay claim but can do so if he does not have a GRC (and vice versa). That is an odd divergence and is unlikely to have been intended by Parliament. It is also true that a biological definition of sex would transfer this right from some trans women to some trans men. We do not see this difficulty as compelling a different conclusion in these circumstances.

263. The anomaly for trans people is a consequence of the requirement in section 64(1)(a) of the EA 2010 to identify an actual comparator of the opposite sex in order to bring an equal pay claim. But, since on either definition of sex, some trans people will not be able to use the equal pay route because of the express requirement for a comparator of the opposite sex, we do not regard this anomaly as mandating a different conclusion.

(21) Summary on the EA 2010

264. For all these reasons, this examination of the language of the EA 2010, its context and purpose, demonstrate that the words "sex", "woman" and "man" in sections 11 and 212(1) mean (and were always intended to mean) biological sex, biological woman and biological man. These and the other provisions to which we have referred cannot properly be interpreted as also extending to include certificated sex without rendering them incoherent and unworkable. In other words, in relation to sex discrimination (for the purposes of sections 11 and 212(1)), a person with the protected characteristic of sex has the characteristic of their biological sex only: a trans man with a GRC (a biological female but legally male for those purposes to which section 9(1) of the GRA 2004 applies) is a woman for the purposes of section 11 and a trans woman with a GRC (biologically male but legally female for those purposes to which section 9(1) applies), is a man and not entitled to be treated as a woman under the EA 2010. This conclusion does not remove or diminish the important protections available under the EA 2010 for trans people with a GRC as we have explained. To the contrary, this potentially vulnerable group remains protected in the ways we have described. In these circumstances, and notwithstanding that there is no express provision in the EA 2010 addressing the effect which section 9(1) of the GRA 2004 has on the definition of "sex", we are satisfied that the EA 2010 does make provision within the meaning of section 9(3) that disapplies the rule in section 9(1) of the GRA 2004.

(22) Summary of our reasoning

265. We are aware that this is a long judgment. It may assist therefore if we summarise our reasoning.

(i) The question for the court is a question of statutory interpretation; we are concerned with the meaning of the provisions of the EA 2010 in the light of section 9 of the GRA (para 2).

(ii) Parliament in using the words "man" and "woman" in the SDA 1975 referred to biological sex (paras 36-51).

(iii) The 1999 Regulations, enacted in response to *P v S*, created a new protected characteristic of a person intending to undergo, or undergoing or having undergone gender reassignment. The 1999 Regulations did not amend the meaning of "man" or "woman" in the SDA 1975 (paras 54-62).

(iv) The GRA 2004 did not amend the meaning of "man" and "woman" in the SDA 1975 (para 80).

(v) Section 9(3) of the GRA 2004 disapplies the rule in section 9(1) of that Act where the words of legislation, enacted before or after the commencement of the GRA 2004, are on careful consideration interpreted in their context and having regard to their purpose to be inconsistent with that rule. It is not necessary that there are express words disapplying the rule in section 9(1) of the GRA 2004 or that such disapplication arises by necessary implication as the legality principle does not apply (paras 99-104).

(vi) The context in which the EA 2010 was enacted was therefore that the SDA 1975 definitions of "man" and "woman" referred to biological sex and trans people had the protected characteristic of gender reassignment.

(vii) The EA 2010 is an amending and consolidating statute. It enacts group-based protections against discrimination on the grounds of sex and gender reassignment and imposes duties of positive action (paras 113, 142-149).

(viii) It is important that the EA 2010 is interpreted in a clear and consistent way so that groups which share a protected characteristic can be identified by those on whom the Act imposes obligations so that they can perform those obligations in a practical way (paras 151-154).

(ix) There is no indication in relevant secondary materials that the EA 2010 modified in any material way the meaning of "man" and "woman" or "sex" from the meanings in the SDA 1975 (para 164).

(x) Interpreting "sex" as certificated sex would cut across the definitions of "man" and "woman" and thus the protected characteristic of sex in an incoherent way. It would create heterogeneous groupings. As a matter of ordinary language, the provisions relating to sex discrimination, and especially those relating to pregnancy and maternity (sections 13(6), 17 and 18), and to protection from risks specifically affecting women (Schedule 22, paragraph 2), can only be interpreted as referring to biological sex (paras 172, 177-188).

(xi) We reject the suggestion of the Inner House that the words can bear a variable meaning so that in the provisions relating to pregnancy and maternity the EA 2010 is referring to biological sex only, while elsewhere it refers to certificated sex as well (paras 189-197).

(xii) Gender reassignment and sex are separate bases for discrimination and inequality. The interpretation favoured by the EHRC and the Scottish Ministers would create two sub-groups within those who share the protected characteristic of gender reassignment, giving trans persons who possess a GRC greater rights than those who do not. Those seeking to perform their obligations under the Act would have no obvious means of distinguishing between the two sub-groups to whom different duties were owed, particularly since they could not ask persons whether they had obtained a GRC (paras 198-203).

(xiii) That interpretation would also seriously weaken the protections given to those with the protected characteristic of sexual orientation for example by interfering with their ability to have lesbian-only spaces and associations (paras 204-209).

(xiv) There are other provisions whose proper functioning requires a biological interpretation of "sex". These include separate spaces and single-sex services (including changing rooms, hostels and medical services), communal accommodation and others (paras 210-228).

(xv) Similar incoherence and impracticability arise in the operations of provisions relating to single-sex characteristic associations and charities, women's fair participation in sport, the operation of the public sector equality duty, and the armed forces (paras 229-246).

(xvi) It is striking that the EHRC has advised the UK Government of the problems created by its interpretation of the EA 2010, which include many of the matters which we have discussed above, and has called for legislation to amend the Act. The absence of coherence and the practical problems to which that interpretation gives rise are clear pointers that the interpretation is not correct (para 247).

(xvii) The interpretation of the EA 2010 (ie the biological sex reading), which we conclude is the only correct one, does not cause disadvantage to trans people, with or without a GRC. In the light of case law interpreting the relevant provisions, they would be able to invoke the provisions on direct discrimination and harassment, and indirect discrimination. A certificated sex reading is not required to give them those protections (paras 248-263).

(xviii) We therefore conclude that the provisions of the EA 2010 which we have discussed are provisions to which section 9(3) of the GRA 2004 applies.

The meaning of the terms "sex", "man" and "woman" in the EA 2010 is biological and not certificated sex. Any other interpretation would render the EA 2010 incoherent and impracticable to operate (para 264).

(23) Invalidity of the Scottish Government's Guidance

266. For all these reasons, we conclude that the Guidance issued by the Scottish Government is incorrect. A person with a GRC in the female gender does not come within the definition of "woman" for the purposes of sex discrimination in section 11 of the EA 2010. That in turn means that the definition of "woman" in section 2 of the 2018 Act, which Scottish Ministers accept must bear the same meaning as the term "woman" in section 11 and section 212 of the EA 2010, is limited to biological women and does not include trans women with a GRC. Because it is so limited, the 2018 Act does not stray beyond the exception permitted in section L2 of Schedule 5 to the Scotland Act into reserved matters. Therefore, construed in the way that we have held it is to be construed, the 2018 Act is within the competence of the Scottish Parliament and can operate to encourage the participation of women in senior positions in public life.

267. There may well be public boards on which it is also important for trans people of either or both genders to be represented in order to ensure that their perspective is brought to bear in the board's deliberations and in the organisation's governance. Nothing in this judgment is intended to discourage the appointment of trans people to public boards or to minimise the importance of addressing their under-representation on such boards. The issue here is only whether the appointment of a trans woman who has a GRC counts as the appointment of a woman and so counts towards achieving the goal set in the gender representation objective, namely that the board has 50% of non-executive members who are women. In our judgment it does not.

(24) Conclusion

268. We would allow the appeal.

End of U.K. Judgment on Definition of a Woman.

NOTES

1. Baumle and Nordmarken, p. 76.
2. Savic, p. 67.
3. Website: www.nytimes.com/2025/02/06/us/politics/ncaa-transgender-athletes-ban.html.
4. For the full text of this Executive Order, see Appendix A.
5. Some might suggest that this "female privilege" has something to do with the fact that the discus throw is currently one of the few physical (as opposed to cerebral) dominant sports in which a woman has outperformed a man: As of this writing the men's discus throw world record is 243 feet 11 inches; the current women's discus throw world record is 251 feet 11 inches.
6. Let us note here that because technique, agility, and flexibility, are more important in rock climbing than height and strength, this particular sport is not as dependent on sex-based divisions as others sports.
7. E. O. Wilson, p, 4. (Note: This is a slight paraphrasal of Wilson's original statement. L.S.)
8. See Johanson and Edgar, p. 73; Steele and Shennan, passim; Dahlberg, pp. 97-98.
9. It is a fact that in some cultures women have assisted men in big game hunting, or even hunted big game by themselves. See e.g., Ellanna and Burch, p. 12. However, these practices are relatively rare among human societies. Based on anthropological and archaeological findings, as well as studies of living archaic peoples, by and large the traditional view of early men as hunters and early women as gatherers holds true.
10. See Short and Balaban, p. 162. Darwin came up with a similar hypothesis. See Hoquet, p. 12.
11. I have coined the word androarchy to mean a "male-centered, male-ruled brotherhood."
12. I have coined the word gynoarchy to mean a "female-centered, female-ruled sisterhood."
13. For more on the biology behind human behavior, and in particular human relationships, see, for example, E. O. Wilson's classic work on evolutionary psychology, *Sociobiology: The New Synthesis*. See also Desmond Morris' classic work, *The Naked Ape: A Zoologist's Study of the Human Animal*.
14. What caused sexual dimorphism to develop in *Homo sapiens* to begin with? Many theories have been put forth, from natural selection to sexual selection (by females), from food niche partitioning to male protection and defensiveness, from habitat quality to foraging behavior, from male competition for females to polygamous social structures, from anisogamy to fecundity, etc. (For example, see Lee, pp. 355-357; Jungers, p. 53; King, p. 40; Hochberg and Campbell, pp. 119-120; Schaik, pp. 144-155.) While fascinating this topic is beyond the scope of the present book.
15. Hochberg and Campbell, p. 119.
16. On the topic of canine teeth, see Oxnard, p. 14-17.
17. Gray, pp. 241-242.
18. For an in-depth look at the evolutionary factors behind the development of the human pelvis, see Wall-Scheffler, Kurki, and Auerbach.
19. We tie with the gorilla and the white-faced sakis. Buss, p. 391.
20. Beat Knechtle, Fabio Valeri, Pantelis T. Nikolaidis, Matthias A. Zingg, Thomas Rosemann, and Christoph A. Rüst: "Do Women Reduce the Gap to Men in Ultra-Marathon Running?" SpringerPlus, May 20, 2016.
21. FOX News, "Jesse Watters Primetime," December 20, 2024.
22. Department of Justice Press Conference, Washington, D.C., April 16, 2025.
23. Department of Justice Press Conference, Washington, D.C., April 16, 2025.
24. FOX News, "FOX Sports," November 9, 2023.
25. Michael Rudling, *Daily Mail*, June 2, 2023.
26. See Dixson, *Sexual Selection and the Origins of Human Mating Systems*, pp. 125-126.
27. For an in-depth treatment of this topic, see Matsumoto, passim.
28. Lanzenberger, Kranz, and Savic, p. 4.

29. The statements that follow, though factual, are generalized and are not meant to be scientifically complete or literal in every sense. For those interested in learning more about these subjects, please refer to my bibliography for suggested reading.

30. Baron-Cohen, pp. 177-178.

31. Savic, pp. 4-5.

32. Savic, pp. 4-5.

33. Baron-Cohen, p. 175.

34. Savic, p. 6.

35. See Baron-Cohen, passim.

36. Neill and Kulkarni, pp. 232-238.

37. Savic, p. 67.

38. Savic, p. 67.

39. The link between inventiveness and hunting is well-known among the knowledgeable. Even the left-wing *Encyclopedia Britannica* recognizes that the two go hand in hand. See Website: https://www.britannica.com/sports/hunting-sport.

40. For more on these topics, see Meghan Van Zandt, Deirdre Flanagan, and Christopher Pittenger, "Sexual Dimorphism in the Distribution and Density of Regulatory Interneurons in the Striatum," passim, National Library of Medicine (PubMed), April 9, 2024.

41. Strier, p. 118.

42. Bruce Goldman, "Two Minds: The Cognitive Differences Between Men and Women," *Stanford Medicine Magazine*, May 22, 2017.

43. "Gaines for Girls," Outkick YouTube channel, March 2025.

44. Genesis 1:27.

45. Matthew 19:1-6. See also Genesis 2:24.

46. For more on these topics see my many books on religion and spirituality in the bibliography.

47. In compiling these lists I acknowledge the fact that the ancients also sometimes associated men with the earth and the moon, and women with the heavens and the sun. My lists, however, represent the ancient norm of connecting men with yang energy and women with yin energy.

48. The symbolism of the Star of David has numerous other fascinating and meaningful interpretations that are outside the scope of this book.

49. The symbolism of the Christian Cross has numerous other fascinating and meaningful interpretations that are outside the scope of this book.

50. Seabrook, *Seabrook's Bible Dictionary of Traditional and Mystical Christian Doctrines*, s.v. "Paul," and s.v. "Cana Wedding."

51. Robinson, pp. 152-153.

52. Tragically for men, over time lawfare-minded misandrists have converted this once biologically-appropriate court procedure into a feminist one. On the positive side, however, the weaponization of the courts against men has created serious backlash, with judges now awarding child custody to fathers more than ever before—especially in cases where the mother is irresponsible, abusive, or drug or alcohol addicted.

53. This may be possible due to the fact that our brains, both male and female, are ambitypic. See Greenberg, Bruess, and Conklin, p. 250.

54. Hoquet, p. 275.

55. See E. O. Wilson, p. 568.

56. For more on this feminist-socialist topic, see Russell, passim.

57. See E. O. Wilson, pp. 36, 46, 314, 328, 332-335, 339, 524, 529, 595.

58. Let us note that sexual dimorphism works both ways. In some animal species, such as blue whales, Bengal floricans, triplewart seadevil anglerfish, gray whales, frigate birds, garter snakes, humpback whales, horned lizards, baleen whales, and spotted hyenas, we find reverse sexual dimorphism (also known as female-biased sexual dimorphism), in which females are larger than males. There are other types of sexual dimorphism as well, such as sexual dichromatism (differences in coloration), found in primates like the blue-eyed black lemur. When there is no apparent sexual dimorphism between the two sexes of the same species, it is called sexual monomorphism—the rarest condition in the animal kingdom. Some examples: the corvid family, chickadees, swans, egrets, sandpipers, and vultures. These species can only be distinguished by examining the genitalia.

59. Carpenter, pp. 277-298; Sullivan, Lucas, and Spielmann, pp. 289-294.

60. Stella A. Ludwig, Roy E. Smith, and Nizar Ibrahim, "Palaeontology: Sexual Dimorphism in Dinosaurs," *eLife Sciences*, June 14, 2023. See also, Romain Pintore, Raphaël Cornette, Alexandra Houssaye, and Ronan Allain, "Femora From an Exceptionally Large Population of Coeval Ornithomimosaurs Yield Evidence of Sexual Dimorphism in Extinct Theropod Dinosaurs," *eLife Sciences*, June 13, 2023.

61. Fairbairn, Blanckenhorn, and Székely, pp. 16, 25.

62. Ruff, p. 289.

63. Orgogozo, p. 378. Naturally this view is debated in the scientific community. See, e.g., B. D. Miller, pp. 46-48.

64. Cartwright, pp. 85-86.

65. Orgogozo, p. 378.

66. I would have liked to add the Denisovans (*Homo denisova*) to my list, but currently there is not enough fossil evidence to determine comparative height, weight, and cranial capacity for Denisovan males and females.

67. Cro-Magnons are more commonly classified today as "anatomically modern humans" (AMH), "European early modern humans" (EEMH), or "early modern humans" (EMH).

68. Due to the current scarcity of crania and postcrania skeletons of *Homo erectus* and *Homo habilis*, I was not able to procure accurate brain size estimations for these two species. However, if I were to make an educated guess, I would say that the brains of both *erectus* and *habilis* males were larger, perhaps much larger, than the brains of their female counterparts. For more on the brain capacity of early hominids, see Papani, pp. 455-493.

69. Orgogozo, p. 378.

70. Orgogozo, p. 378.

71. For more on this topic, see Tuttle, pp. 245-281; Reed, Fleagle, and Leakey, pp. 195-212.

72. Johanson and Edgar, p. 73.

73. King, p. 41.

74. Oxnard, p. 13.

75. Orgogozo, p. 378.

76. See Johanson and Edgar, p. 73.

77. See the Sex Matters Website: https://www.sex-matters.org

78. Asher Notheis, "Riley Gaines Joins Virginia Women's Swim Team to Advocate Against Transgender Athletes," *Washington Examiner*, October 25, 2023.

79. Ben Talintyre, "Riley Gaines Speaks Out Amid Donald Trump's Historic Transgender Sport Ban," News.com.au, February 28, 2025.

80. I, of course, recognize the fact that some individuals are born with an extra or missing Y or X chromosome—such as we find in an XXY or an XXX chromosomal sex pattern. I also understand that some people experience biological changes in their genes that categorize them as intersex. But as both are considered "genetic conditions," and in some cases, "genetic disorders," such rare exceptions to the genetic binary rule are not germane to the topic of this book. Note: Despite extraordinary claims to the contrary, only 0.018% of the population is considered intersex. See Leonard Sax, "How Common is Intersex? A Response to Anne Fausto-Sterling," National Library of Medicine (NIH), National Center for Biotechnology Information. PubMed Website: https://pubmed.ncbi.nlm.nih.gov/12476264/.

81. Department of Justice Press Conference, Washington, D.C., April 16, 2025.

82. Baumle and Nordmarken, p. 76.

All-women volleyball match, 1927.

Horsewoman Miss Ina Gittings taking a jump, Tucson, Arizona, 1927.

BIBLIOGRAPHY

And Suggested Reading

Abbot, Senda Berenson (ed.). *Spaulding's Athletic Library: Official Basketball Guide for Women, 1916-1917*. New York: American Sports Publishing Co., 1916.

Alexander, Linda Lewis, Judith H. LaRosa, Helaine Bader, and Susan Garfield. *New Dimensions In Women's Health*. Sudbury, MA: Jones and Bartlett, 2010.

American Psychiatric Association. *Diagnostic and Statistical Manual of Mental Disorders* (5th ed., DSM-5). Washington, D.C.: American Psychiatric Publishing, 2017.

Applebee, Constance Mary Katherine (pub). *The Sportswoman*. Issues 1926-1928. U.S.A.

Ardrey, Robert. *African Genesis*. 1961. New York: Dell, 1972 ed.

——. *The Territorial Imperative*. 1966. New York: Delta, 1968 ed.

Armstrong, Heather L. (ed.). *Encyclopedia of Sex and Sexuality: Understanding Biology, Psychology, and Culture*. London, UK: Bloomsbury Publishing, 2021.

Arnqvist, Göran, and Locke Rowe. *Sexual Conflict*. Princeton, NJ: Princeton University Press, 2005.

Baikie, James. *Peeps at Men of the Old Stone Age*. London, UK: A. and C. Black, 1928.

Barnard, Alan. *Hunter-Gatherers in History, Archaeology and Anthropology*. 2004. Abingdon, UK: Routledge, 2020 ed.

Baron-Cohen, Simon. *The Essential Difference: Male and Female Brains and the Truth about Autism*. New York: Perseus Books Group, 2003.

Baumle, Amanda K., and Sonny Nordmarken (eds.). *Demography of Transgender, Nonbinary and Gender Minority Populations*. Cham, Switzerland: Springer Nature, 2022.

Begun, David R. (ed.). *A Companion to Paleoanthropology*. Hoboken, NJ: John Wiley and Sons, 2012.

Burch, Ernest S., Jr., and Linda J. Ellanna (eds.). *Key Issues in Hunter-Gatherer Research*. 1994. Abingdon, UK: Routledge, 2020 ed.

Buss, David M. *The Handbook of Evolutionary Psychology, Volume 1: Foundations*. Hoboken, NJ: John Wiley and Sons, 2016.

Carpenter, Kenneth (ed.). *The Carnivorous Dinosaurs*. Bloomington, IN: Indiana University Press, 2005.

Cartwright, John. *Evolution and Human Behaviour: Darwinian Perspectives on the Human Condition*. 2000. London, UK: Palgrave, 2016 ed.

Choe, Jae Chun (ed.). *Encyclopedia of Animal Behavior*. Amsterdam, The Netherlands: Elsevier, 2019.

Cummings, Vicki. *The Anthropology of Hunter-Gatherers: Key Themes for Archaeologists*. 2013. Abingdon, UK: Routledge, 2020 ed.

Dahlberg, Frances (ed.). *Woman the Gatherer*. New Haven, CT: Yale University Press, 1981.

Dirkmaat, Dennis C. *A Companion to Forensic Anthropology*. 2012. Hoboken, NJ: John Wiley and Sons, 2015 ed.

Dixson, Alan F. *Primate Sexuality: Comparative Studies of the Prosimians, Monkeys, Apes, and Humans*. 1998. Oxford, UK: Oxford University Press, 2012 ed.

——. *Sexual Selection and the Origins of Human Mating Systems*. Oxford, UK: Oxford University Press, 2009.

Ellis, Havelock. *Man and Woman: A Study of Secondary and Tertiary Sexual Characters*. Boston, MA: Houghton Mifflin Co., 1929.

Fairbairn, Daphne J., Wolf U. Blanckenhorn, and Tamás Székely (eds.). *Sex, Size and Gender Roles: Evolutionary Studies of Sexual Size Dimorphism*. Oxford, UK: Oxford University Press, 2007.

Fedigan, Linda Marie. *Primate Paradigms: Sex Roles and Social Bonds*. 1982. Chicago, IL: University of Chicago Press, 1992 ed.

Fisher, Maryanne L. *The Oxford Handbook of Women and Competition*. Oxford, UK: Oxford University Press, 2017.

Fortin, Timothy Paul. *On the Nature of Human Sexual Difference: A Symposium*. Cham, Switzerland: Springer Nature, 2024.

Furtwängler, Adolf, and Heinrich Ludwig Urlichs. *Greek and Roman Sculpture*. London, UK: J. M Dent and Sons, 1914.

Gray, Henry. *Anatomy of the Human Body*. Philadelphia, PA: Lea and Febiger, 1918.

Greenberg, Jerrold, Clint Bruess, and Sarah Conklin. *Exploring the Dimensions of Human Sexuality*. Sudbury, MA: Jones and Bartlett, 2011.

Hall, Roberta L. *Sexual Dimorphism in Homo Sapiens: A Question of Size*. Westport, CT: Praeger, 1982.

Hochberg, Zeev, and Benjamin C. Campbell (eds.). *Evolutionary Perspectives on Human Growth and Development*. Lausanne, Switzerland: Frontiers Media, 2021.

Hoquet, Thierry (ed.). *Current Perspectives on Sexual Selection: What's Left After Darwin?* Dordrecht, The Netherlands: Springer, 2015.

Johanson, Donald. *Lucy: The Beginnings of Humankind*. New York, NY: Touchstone, 1981.

Johansan, Donald, and Blake Edgar. *From Lucy to Language*. New York: Simon and Schuster, 1996.

Junger, William L. (ed.). *Size and Scaling in Primate Biology*. New York: Springer Science, 1985.

Kelly, Robert L. *The Lifeways of Hunter-Gatherers: The Foraging Spectrum*. 1995. Cambridge, UK: Cambridge University Press, 2013 ed.

King, Glenn E. *Baboon Perspectives on Early Human Ancestors: One Approach to Reconstructing Early Hominin Ecology and Behavior*. Cham, Switzerland: Springer Nature, 2024.

Lancaster, Jane Beckman. *Primate Behavior and the Emergence of Human Culture*. New York: Holt, Rinehart and Winston, 1975.

Lanzenberger, Rupert, Georg S. Kranz, and Ivanka Savic (eds.). *Sex Differences in Neurology and Psychiatry*. Amsterdam, The Netherlands: Elsevier, 2020.

Leakey, Richard E., and Roger Lewin. *Origins Reconsidered: In Search of What Makes Us Human*. New York, NY: Doubleday, 1992.

Lee, P. C. (ed.). *Comparative Primate Socioecology*. Cambridge, UK: Cambridge University Press, 1999.

Lydekker, Richard. *Wild Life of the World: A Descriptive Survey of the Geographical Distribution of Animals*. London, UK: Frederick Warne and Co., 1916.

Mach, Edmund von. *Greek Sculpture: Its Spirit and Principles*. Boston, MA: Ginn and Co., 1903.

Mackey, Wade C., and Nancy S. Coney. *Gender Roles, Traditions, and Generations to Come: The Collision of Competing Interests and the Feminist Paradox*. Huntington, NY: Nova Science, 2000.

Matsumoto, Akira (ed.). *Sexual Differentiation of the Brain*. Boca Raton, FL: CRC Press, 1999.

Miller, Barbara Diane (ed.). *Sex and Gender Hierarchies*. Cambridge, UK: Cambridge University Press, 1993.

Miller, Geoffrey. *The Mating Mind: How Sexual Choice Shaped the Evolution of Human Nature*. New York: Anchor Books, 2000.

Neill, Jo C., and Jayashri Kulkarni (eds.). *Biological Basis of Sex Differences in*

Psychopharmacology. Heidelberg, Germany: Springer-Verlag, 2011.

Orgogozo, Virginie (ed.). *Current Topics in Developmental Biology: Genes and Evolution.* Amsterdam, The Netherlands: Elsevier, 2016.

Osburn, Henry Fairfield. *Men of the Old Stone Age: Their Environment, Life and Art.* New York: Charles Scribner's Sons, 1915.

Oxnard, Charles E. *Human, Apes, and Chinese Fossils: New Implications for Human Evolution.* Hong Kong, China: Hong Kong University Press, 1985.

Papini, Mauricio R. *Comparative Psychology: Evolution and Development of Brain and Behavior.* 2001. New York: Routledge, 3rd ed., 2021.

Reed, Kaye E., John G. Fleagle, and Richard E. Leakey (ed.). *The Paleobiology of Australopithecus.* Dordrecht, The Netherlands: Springer, 2013.

Robinson, Paschal. *The Writings of Saint Francis of Assisi.* Philadelphia, PA: The Dolphin Press, 1905.

Ross, Callum F., and Richard F. Kay (eds.). *Anthropoid Origins: New Visions.* New York: Springer Science, 2004.

Ruff, Christopher B. (ed.). *Skeletal Variation and Adaptation in Europeans: Upper Paleolithic to the Twentieth Century.* Hoboken, NJ: John Wiley and Sons, 2018.

Sassaman, Kenneth E., and Donald H. Holly, Jr. (eds.). *Hunter-gatherer Archaeology as Historical Process.* Tucson, AZ: University of Arizona Press, 2011.

Savic, Ivanka (ed.). *Sex Differences in the Human Brain, Their Underpinnings and Implications.* Amsterdam, The Netherlands: Elsevier, 2010.

Schaik, Carel P. Van. *The Primate Origins of Human Nature.* Hoboken, NJ: John Wiley and Sons, 2016.

Seabrook, Lochlainn. *Aphrodite's Trade: The Hidden History of Prostitution Unveiled.* 1994. Franklin, TN: Sea Raven Press, 2011 ed.

——. *The Goddess Dictionary of Words and Phrases: Introducing a New Core Vocabulary for the Women's Spirituality Movement.* 1997. Franklin, TN: Sea Raven Press, 2010 ed.

——. *Britannia Rules: Goddess-Worship in Ancient Anglo-Celtic Society - An Academic Look at the United Kingdom's Matricentric Spiritual Past.* 1999. Franklin, TN: Sea Raven Press, 2010 ed.

——. *The Book of Kelle: An Introduction to Goddess-Worship and the Great Celtic Mother-Goddess Kelle, Original Blessed Lady of Ireland.* 1999. Franklin, TN: Sea Raven Press, 2010 ed.

——. *Carnton Plantation Ghost Stories: True Tales of the Unexplained from Tennessee's Most Haunted Civil War House!* 2005. Franklin, TN, 2016 ed.

——. *Nathan Bedford Forrest: Southern Hero, American Patriot.* 2007. Franklin, TN, 2010 ed.

——. *Abraham Lincoln: The Southern View.* 2007. Franklin, TN: Sea Raven Press, 2013 ed.

——. *The McGavocks of Carnton Plantation: A Southern History - Celebrating One of Dixie's Most Noble Confederate Families and Their Tennessee Home.* 2008. Franklin, TN, 2011 ed.

——. *Christmas Before Christianity: How the Birthday of the "Sun" Became the Birthday of the "Son."* Franklin, TN: Sea Raven Press, 2010.

——. *A Rebel Born: A Defense of Nathan Bedford Forrest.* 2010. Franklin, TN: Sea Raven Press, 2011 ed.

——. *Everything You Were Taught About the Civil War is Wrong, Ask a Southerner!* 2010. Franklin, TN: Sea Raven Press, 2024 ed.

——. *The Quotable Jefferson Davis: Selections From the Writings and Speeches of the Confederacy's First President.* Franklin, TN: Sea Raven Press, 2011.

——. *The Quotable Robert E. Lee: Selections From the Writings and Speeches of the South's Most Beloved Civil War General.* Franklin, TN: Sea Raven Press, 2011 Sesquicentennial Civil War Edition.

——. *Lincolnology: The Real Abraham Lincoln Revealed In His Own Words.* Franklin, TN: Sea

Raven Press, 2011.

——. *The Unquotable Abraham Lincoln: The President's Quotes They Don't Want You To Know!* Franklin, TN: Sea Raven Press, 2011.

——. *Honest Jeff and Dishonest Abe: A Southern Children's Guide to the Civil War.* Franklin, TN: Sea Raven Press, 2012.

——. *Encyclopedia of the Battle of Franklin - A Comprehensive Guide to the Conflict that Changed the Civil War.* Franklin, TN: Sea Raven Press, 2012.

——. *The Quotable Nathan Bedford Forrest: Selections From the Writings and Speeches of the Confederacy's Most Brilliant Cavalryman.* Spring Hill, TN: Sea Raven Press, 2012.

——. *Forrest! 99 Reasons to Love Nathan Bedford Forrest.* Spring Hill, TN: Sea Raven Press, 2012.

——. *Give 'Em Hell Boys! The Complete Military Correspondence of Nathan Bedford Forrest.* Spring Hill, TN: Sea Raven Press, 2012.

——. *The Constitution of the Confederate States of America Explained: A Clause-by-Clause Study of the South's Magna Carta.* Spring Hill, TN: Sea Raven Press, 2012 Sesquicentennial Civil War Edition.

——. *The Great Impersonator: 99 Reasons to Dislike Abraham Lincoln.* Spring Hill, TN: Sea Raven Press, 2012.

——. *The Old Rebel: Robert E. Lee As He Was Seen By His Contemporaries.* Spring Hill, TN: Sea Raven Press, 2012 Sesquicentennial Civil War Edition.

——. *The Quotable Stonewall Jackson: Selections From the Writings and Speeches of the South's Most Famous General.* Spring Hill, TN: Sea Raven Press, 2012 Sesquicentennial Civil War Edition.

——. *Saddle, Sword, and Gun: A Biography of Nathan Bedford Forrest for Teens.* Spring Hill, TN: Sea Raven Press, 2013.

——. *Jesus and the Law of Attraction: The Bible-Based Guide to Creating Perfect Health, Wealth, and Happiness Following Christ's Simple Formula.* Franklin, TN: Sea Raven Press, 2013.

——. *The Bible and the Law of Attraction: 99 Teachings of Jesus, the Apostles, and the Prophets.* Franklin, TN: Sea Raven Press, 2013.

——. *The Alexander H. Stephens Reader: Excerpts From the Works of a Confederate Founding Father.* Spring Hill, TN: Sea Raven Press, 2013.

——. *The Quotable Alexander H. Stephens: Selections From the Writings and Speeches of the Confederacy's First Vice President.* Spring Hill, TN: Sea Raven Press, 2013 Sesquicentennial Civil War Edition.

——. *Christ Is All and In All: Rediscovering Your Divine Nature and the Kingdom Within.* Franklin, TN: Sea Raven Press, 2014.

——. *Jesus and the Gospel of Q: Christ's Pre-Christian Teachings as Recorded in the New Testament.* Franklin, TN: Sea Raven Press, 2014.

——. *Give This Book to a Yankee! A Southern Guide to the Civil War for Northerners.* Spring Hill, TN: Sea Raven Press, 2014.

——. *The Articles of Confederation Explained: A Clause-by-Clause Study of America's First Constitution.* Spring Hill, TN: Sea Raven Press, 2014.

——. *Confederate Blood and Treasure: An Interview With Lochlainn Seabrook.* Spring Hill, TN: Sea Raven Press, 2015.

——. *Nathan Bedford Forrest and the Battle of Fort Pillow: Yankee Myth, Confederate Fact.* Spring Hill, TN: Sea Raven Press, 2015.

——. *Everything You Were Taught About American Slavery War is Wrong, Ask a Southerner!* Spring Hill, TN: Sea Raven Press, 2015.

——. *Confederacy 101: Amazing Facts You Never Knew About America's Oldest Political Tradition.* Spring Hill, TN: Sea Raven Press, 2015.

——. *The Great Yankee Coverup: What the North Doesn't Want You to Know About Lincoln's War!*

Spring Hill, TN: Sea Raven Press, 2015.

——. *Slavery 101: Amazing Facts You Never Knew About America's "Peculiar Institution."* Spring Hill, TN: Sea Raven Press, 2015.

——. *Confederate Flag Facts: What Every American Should Know About Dixie's Southern Cross.* Spring Hill, TN: Sea Raven Press, 2016.

——. *Nathan Bedford Forrest and the Ku Klux Klan: Yankee Myth, Confederate Fact.* Spring Hill, TN: Sea Raven Press, 2016.

——. *Seabrook's Bible Dictionary of Traditional and Mystical Christian Doctrines.* Spring Hill, TN: Sea Raven Press, 2016.

——. *Everything You Were Taught About African-Americans and the Civil War is Wrong, Ask a Southerner!* Spring Hill, TN: Sea Raven Press, 2016.

——. *Nathan Bedford Forrest and African-Americans: Yankee Myth, Confederate Fact.* Spring Hill, TN: Sea Raven Press, 2016.

——. *Women in Gray: A Tribute to the Ladies Who Supported the Southern Confederacy.* Spring Hill, TN: Sea Raven Press, 2016.

——. *Lincoln's War: The Real Cause, the Real Winner, the Real Loser.* Spring Hill, TN: Sea Raven Press, 2016.

——. *The Unholy Crusade: Lincoln's Legacy of Destruction in the American South.* Spring Hill, TN: Sea Raven Press, 2017.

——. *Abraham Lincoln Was a Liberal, Jefferson Davis Was a Conservative: The Missing Key to Understanding the American Civil War.* Spring Hill, TN: Sea Raven Press, 2017.

——. *All We Ask is to be Let Alone: The Southern Secession Fact Book.* Spring Hill, TN: Sea Raven Press, 2017.

——. *The Ultimate Civil War Quiz Book: How Much Do You Really Know About America's Most Misunderstood Conflict?* Spring Hill, TN: Sea Raven Press, 2017.

——. *Rise Up and Call Them Blessed: Victorian Tributes to the Confederate Soldier, 1861-1901.* Spring Hill, TN: Sea Raven Press, 2017.

——. *Victorian Confederate Poetry: The Southern Cause in Verse, 1861-1901.* Spring Hill, TN: Sea Raven Press, 2018.

——. *Confederate Monuments: Why Every American Should Honor Confederate Soldiers and Their Memorials.* Spring Hill, TN: Sea Raven Press, 2018.

——. *The God of War: Nathan Bedford Forrest as He Was Seen by His Contemporaries.* Spring Hill, TN: Sea Raven Press, 2018.

——. *The Battle of Spring Hill: Recollections of Confederate and Union Soldiers.* Spring Hill, TN: Sea Raven Press, 2018.

——. *I Rode With Forrest! Confederate Soldiers Who Served With the World's Greatest Cavalry Leader.* Spring Hill, TN: Sea Raven Press, 2018.

——. *The Battle of Nashville: Recollections of Confederate and Union Soldiers.* Spring Hill, TN: Sea Raven Press, 2018.

——. *The Battle of Franklin: Recollections of Confederate and Union Soldiers.* Spring Hill, TN: Sea Raven Press, 2018.

——. *A Rebel Born: The Screenplay* (for the film). Written 2011. Franklin, TN: Sea Raven Press, 2020.

——. (ed.) *A Short History of the Confederate States of America* (Jefferson Davis, Belford Company, NY, 1890). A Sea Raven Press Reprint. Spring Hill, TN: Sea Raven Press, 2020.

——. (ed.) *Prison Life of Jefferson Davis: Embracing Details and Incidents in his Captivity, With Conversations on Topics of Public Interest* (John J. Craven, Sampson, Low, Son, and Marston, London, UK, 1866). A Sea Raven Press Reprint. Spring Hill, TN: Sea Raven Press, 2020.

——. *What the Confederate Flag Means to Me: Americans Speak Out in Defense of Southern Honor,*

Heritage, and History. Spring Hill, TN: Sea Raven Press, 2021.

——. *Heroes of the Southern Confederacy: The Illustrated Book of Confederate Officials, Soldiers, and Civilians*. Spring Hill, TN: Sea Raven Press, 2021.

——. *Support Your Local Confederate: Wit and Humor in the Southern Confederacy*. Spring Hill, TN: Sea Raven Press, 2021.

——. *America's Three Constitutions: Complete Texts of the Articles of Confederation, Constitution of the United States of America, and Constitution of the Confederate States of America*. Spring Hill, TN: Sea Raven Press, 2021.

——. *Vintage Southern Cookbook: 2,000 Delicious Dishes From Dixie*. Spring Hill, TN: Sea Raven Press, 2021.

——. *The Bittersweet Bond: Race Relations in the Old South as Described by White and Black Southerners*. Spring Hill, TN: Sea Raven Press, 2022.

——. (ed.) *The Rise and Fall of the Confederate Government* (Jefferson Davis, D. Appleton, New York, 1881). 2 vols. A Sea Raven Press Facsimile Reprint. Spring Hill, TN: Sea Raven Press, 2022.

——. *Secrets of Celebrity Surnames: An Onomastic Dictionary of Famous People*. Cody, WY: Sea Raven Press, 2023.

——. *I, Confederate: Why Dixie Seceded and Fought in the Words of Southern Soldiers*. Spring Hill, TN: Sea Raven Press, 2023.

——. *Twelve Years in Hell: Victorian Southerners Expose the Myth of Reconstruction, 1865-1877*. Cody, WY: Sea Raven Press, 2023.

——. *Seabrook's Complete Battle Book: The War Between the States, 1861-1865*. Cody, WY: Sea Raven Press, 2023.

——. *The Hampton Roads Conference: The Southern View*. Cody, WY: Sea Raven Press, 2024.

——. *Rocky Mountain Equines: A Photographic Collection of Horses, Donkeys, and Mules of the American West*. Cody, WY: Sea Raven Press, 2024.

——. *Rocky Mountain Bison: A Photographic Collection of Bison of the American West*. Cody, WY: Sea Raven Press, 2024.

——. *Mysterious Invaders: Twelve Famous 20th-Century Scientists Confront the UFO Phenomenon*. Cody, WY: Sea Raven Press, 2024.

——. *We Called Him Jeb: James Ewell Brown Stuart as He Was Seen by His Contemporaries*. Cody, WY: Sea Raven Press, 2024.

——. *Your Soul Lives Forever: Documented Victorian Case Studies Proving Consciousness Survives Death*. Cody, WY: Sea Raven Press, 2024.

——. *Authentic Victorian Ghost Stories: Genuine Early Reports of Apparitions, Wraiths, Poltergeists, and Haunted Houses*. Cody, WY: Sea Raven Press, 2024.

——. *The Greatest Jesus Mystery of All Time: Where Was Christ Between the Ages of 12 and 30?* Cody, WY: Sea Raven Press, 2024.

——. *Vitamin D: The Miracle Treatment for Nearly Every Disease and Health Issue*. Cody, WY: Sea Raven Press, 2024.

——. *Manmade: Male Inventors Who Created the Modern World*. Cody, WY: Sea Raven Press, 2025.

——. *Jesus and the Gospel of Thomas: A Christian Mystic's View of Christianity's Most Important Ancient Text*. Cody, WY: Sea Raven Press, 2025.

——. *The Way of Holiness: The Story of Religion and Mythology, from the Cave Bear Cult to Christianity: A Study of the Origins, Development, Functions, Symbols, and Themes of Spiritual Thought*. Unpublished manuscript.

——. *Mothers and Bachelors: Ending the Battle of the Sexes—A New Approach to Marriage and the Family Based on the Sciences of Anthropology, Primatology, and Sociobiology*. Unpublished manuscript.

——. *Seabrook's Encyclopedia of Religion and Myth*. Unpublished manuscript.

——. *Families Around the World: A Children's Guidebook to the Marriages and Families of Different Cultures.* Unpublished manuscript.

——. *The True Legend of King Arthur: The Magical Story of Britain's Most Famous Ruler.* Unpublished manuscript.

——. *Glimpses of Heaven: A Guidebook to the Near-Death Experience.* Unpublished manuscript.

Shackleford, Todd K., and Aaron T. Goetz (eds.). *The Oxford Handbook of Sexual Conflict in Humans.* Oxford, UK: Oxford University Press, 2102.

Short, R. V., and E. Balaban (eds.). *The Differences Between the Sexes.* Cambridge, UK: Cambridge University Press, 1994.

Smith, Fred H., and James C. Ahern. *The Origins of Modern Humans: Biology Reconsidered.* Hoboken, NJ: John Wiley and Sons, 2013.

Spurrell, Herbert George Flaxman. *Modern Man and His Forerunners: A Short Study of the Human Species Living and Extinct.* London, UK: G. Bell and Sons, 1917.

Stanford, Craig, John S. Allen, and Susan C. Anton. *Biological Anthropology: The Natural History of Humankind.* Englewood Cliffs, NJ: Prentice Hall, 2005.

Steele, James, and Stephen Shennan (eds.). *The Archaeology of Human Ancestry: Power, Sex, and Tradition.* Abingdon, UK: Routledge, 2017.

Stefoff, Rebecca. *First Humans (Humans: An Evolutionary History).* Tarrytown, NY: Marshall, Cavendish, Benchmark, 2010.

Strier, Karen B. *Primate Behavioral Ecology.* 2000. Abingdon, UK: Routledge, 2011 ed.

Sullivan, Robert M., Spencer G. Lucas, and Justin A. Spielmann (eds.). *Fossil Record.* Albuquerque, NM: New Mexico Museum of Natural History and Science, 2011.

Tuttle, Russell H. (ed.). *Paleoanthropology: Morphology and Paleoecology.* Paris, France: Mouton Publishers, 1975.

United States Dept. of Health and Human Services. *Physical Activity and Sport in the Lives of Girls.* Washington, D.C.: U.S. Government Printing Office, 1997.

Wall-Scheffler, Cara M., Helen K. Kurki, and Benjamin M. Auerbach. *The Evolutionary Biology of the Human Pelvis: An Integrative Approach.* Cambridge, UK: Cambridge University Press, 2020.

Wilson, E. O. *Sociobiology: The New Synthesis.* Cambridge, MA: Harvard University Press, 1975.

Female canoeists, 1920s.

All-women's ice hockey team, Toronto, Canada, 1920s.

College tennis champions, University of Arizona, 1920s.

INDEX

INCLUDES TOPICS, PEOPLE, KEYWORDS, KEY PHRASES, & SPELLING VARIATIONS

MEET THE AUTHOR

A MERICAN POLYMATH LOCHLAINN SEABROOK is a bestselling author, award-winning historian, and world acclaimed artist. A descendant of the families of Alexander Hamilton Stephens, John Singleton Mosby, Edmund Winchester Rucker, and William Giles Harding, the neo-Victorian scholar is a 7th generation Kentuckian, and one of the most prolific and widely read writers in the world today. Known by literary critics as the "new Shelby Foote," the "American Robert Graves," the "Southern Joseph Campbell," and the "Rocky Mountain Richard Jefferies," and by his fans as the "Voice of the Traditional South," he is a recipient of the United Daughters of the Confederacy's prestigious Jefferson Davis Historical Gold Medal, and is considered the foremost Southern interpreter of American Civil War history—or what he refers to as the War for the Constitution (1861-1865). A lifelong litterateur, the Sons of Confederate Veterans member has authored and edited books ranging in topics from ancient and modern history, politics, science, comparative religion, diet and nutrition, spirituality, astronomy, entertainment, military, biography, mysticism, photography, and Bible studies, to natural history, technology, paleography, music, humor, gastronomy, etymology, onomastics, mysteries, alternative health and fitness, wildlife, comparative mythology, genealogy, Christian history, and the paranormal; books that his readers describe as "game changers," "transformative," and "life altering."

One of America's most popular living historians, he is a 17th generation Southerner of Appalachian heritage who descends from dozens of patriotic Revolutionary War soldiers and Confederate soldiers from Kentucky, Tennessee, North Carolina, and Virginia. Also a history, wildlife, and nature preservationist, the well-respected scrivener began life as a child prodigy, later maturing into an archetypal Renaissance Man. Besides being cofounder and co-CEO of Sea Raven Press, an accomplished writer, author, historian, biographer, lexicographer, encyclopedist, neologist, publisher, editor, poet, creative, onomastician, etymologist, and Bible authority, the influential prosateur is also a Kentucky Colonel, eagle scout, entrepreneur, businessman, composer, screenwriter, nature, wildlife, and landscape photographer, videographer, and filmmaker, artist, artisan, painter, watercolorist, sculptor, ceramic artist, visual artist, sketch artist, pen and ink artist, graphic artist, graphic designer, book designer, book formatter, editorial designer, book cover designer, publishing designer, Web designer, poster artist, cartoonist, content creator, inventor, aquarist, genealogist, jewelry designer, jewelry maker, former history museum docent, and a former Red Cross certified lifeguard, ranch hand, zookeeper, and wrangler. A contemporary songwriter (of some 3,000 songs in a dozen genres), he is also a pianist, organist, drummer, bass player, rhythm guitarist, rhythm mandolinist, percussionist, classical composer, film composer (currently his musical work has been featured in 11 movies), lyricist, band leader, multi-instrument musician, lead vocalist, backup vocalist, session player, music producer, and recording studio mixing engineer, who has worked and performed with some of Nashville's top musicians and singers.

Currently Seabrook is the multi-genre author and editor of over 100 adult and children's books (totaling some 30,000 pages and 15,000,000 words) that have earned him accolades from around the globe. His works, which have sold on every continent except Antarctica, have introduced hundreds of thousands to vital facts that have been left out of our mainstream books. He has been endorsed internationally by leading experts, museum curators, award-winning historians, chart-topping authors, celebrities, filmmakers, noted scientists, well regarded educators, TV show hosts and producers, renowned military artists, venerable heritage organizations, and distinguished academicians of all races, creeds, and colors. He currently holds two world records: He is the author of the most books (12) on American military officer Nathan Bedford Forrest, and he was the first to publicize and describe the 19th-Century platform reversal of America's two main political parties, namely that Civil War era Democrats (primarily in the South—the Confederacy) were Conservatives, while Civil War era Republicans (primarily in the North—the Union) were Liberals.

Of northern, western, and central European ancestry, he is the 6th great-grandson of the Earl of Oxford and a descendant of European royalty through his Kentucky father and West Virginia mother. A proud descendant of Appalachian coal miners, trainmen, mountain folk, and wilderness pioneers, his modern day cousins include: Johnny Cash, Elvis Presley, Lisa Marie Presley, Billy Ray and Miley Cyrus, Patty Loveless, Tim McGraw, Lee Ann Womack, Dolly Parton, Pat Boone, Naomi, Wynonna, and Ashley Judd, Ricky Skaggs, the Sunshine Sisters, Martha Carson, Chet Atkins, Patrick J. Buchanan, Cindy Crawford, Bertram Thomas Combs (Kentucky's 50th governor), Edith Bolling (second wife of President Woodrow Wilson), Andy Griffith, Riley Keough, George C. Scott, Robert Duvall, Reese Witherspoon, Lee Marvin, Rebecca Gayheart, and Tom Cruise.

A constitutionalist, avid outdoorsman, wilderness conservationist, and gun rights advocate, Seabrook is the author of the international blockbuster, *Everything You Were Taught About the Civil War is Wrong, Ask a Southerner!* He lives with his wife and family in the magnificent Rocky Mountains, heart of the American West, where you will find him hiking, filming, and writing.

For more information on Mr. Seabrook visit
LOCHLAINNSEABROOK.COM

Praise for Author-Historian-Artist
Lochlainn Seabrook

Comments from our readers around the world

★ "Lochlainn Seabrook is a genius writer!" — STEVEN WARD

★ "Best author ever." — EMILY (last name withheld)

★ "We get asked a lot what books we use and read. We don't do many modern historians, but we make an exception for some, and Lochlainn Seabrook is one of them. His works are completely well researched from original documents, and heavily footnoted and documented." — SOUTHERN HISTORICAL SOCIETY

★ "Looking forward to more Lochlainn Seabrook books, my favourite historian!" — ALBERTO IGLESIAS

★ "Lochlainn Seabrook is one of the finest authors on true history in this century. His books should be on every student's desk." — RONDA SAMMONS RENO

★ "All of Col. Seabrook's books are great. I have bought most of them and want to end up buying them all." — DAVID VAUGHN

★ "Lochlainn pulls together such arcane facts with relative ease, compiling these into ordinary prose that strike to the heart with substance, no fluff-speak. I am awestruck! Really. He is an inspiration to me. . . . He is truly a revolutionist. He dares to speak what others whisper; he writes with a boldness and an authoritative knowledge that is second to none." — JAY KRUIZENGA

★ "Mr. Lochlainn Seabrook is . . . the most well researched and heavily documented author I've ever read. His books are must haves. Everything he writes should be required reading! I assure you, you won't be disappointed. One simply cannot go wrong with his books. Mr. Seabrook is awesome! . . . I have never read any other author as well researched and footnoted as him. I've been in love with Mr. Seabrook for almost 5 years now. His quick wit and logic is enough reason to purchase his books. But the mere fact that he's so extensively researched is icing on the cake. Mr. Seabrook is my favorite, hands down." — LANI BURNETTE RINKEL

★ "My favorite book is the Bible. Lochlainn Seabrook wrote my second favorite book." — RICHARD FINGER

★ "I have a new favorite author and his name is Lochlainn Seabrook." — J. EWING

★ "Lochlainn Seabrook is an incredible writer and I love all of his books on the South. . . . His writing is brilliant. . . . I look forward to reading more of his masterpieces. Thank you." — JOEY (last name withheld)

★ "It's hard to choose just one of Lochlainn's books!" — ROSANNE STEELE

★ "Mr. Seabrook, thank you ever so much for blessing us with your most enlightening works." — LAURENCE DRURY

★ "I recommend anything written by Lochlainn Seabrook." — HOTRODMOB

★ "Awesome books . . . by a great writer of truth, Lochlainn. Thank you so much. Keep up the great work you do." — WILDBUNCH19INF

✯ "I love Lochlainn Seabrook's style and approach. It's not the 'norm.' What a miracle his books are. . . . He is a literal life changing author! Amazing books!" — KEITH PARISH

✯ "I adore Mr. Seabrook's style and I love his books. I love an author that does proper research, and still finds a way to engage the reader. Mr. Seabrook does an admirable job of both." — DONALD CAUL

✯ "Lochlainn Seabrook's books are much more well researched and authoritative than those eminently celebrated as being the authorities on the subjects he writes on. You can always trust to find the truth in his writings. . . . He does not rewrite history, but instead shows it as it is." — GARY STIER

✯ "I love all of Colonel Seabrook's books. They are informative and enlightening, and his warm Southern hospitality writing style makes you feel right at home." — KEITH CRAVEN

✯ "Lochlainn Seabrook's work is an absolute treasure of scholarship and historic scope." — MARK WAYNE CUNNINGHAM

✯ "Mr. Seabrook's command of . . . history is breathtaking. . . . He deserves great renown—check out his books!" — MARGARET SIMMONS

✯ "I love Seabrook's writings. LOVE!!! . . . So grateful to know the truth! Keep writing Lochlainn!!!" — REBECCA DALRYMPLE

✯ "Lochlainn Seabrook . . . [has] probably [written] the best book on mental science in existence by a living author. Along with Thomas Troward, Emmet Fox, and Jack Addington, Mr. Seabrook is one of the top four mental science authors of all time, since biblical times." - IAN BARTON STEWART

✯ "Glad I discovered Mr. Seabrook! . . . He writes eye opening books! Unbelievable the facts he unearths - and he backs it all up with truth, notes, footnotes, and bibliography! . . . He always amazes me! His books always see the whole picture. His timelines and bibliographies are incredible. He always provides carefully reasoned arguments! He's the best. To me I think he's better than the late great Shelby Foote! America needs more like Lochlainn Seabrook. I can't wait to own all of his books on the war someday. Everyone who wants the Truth, who seeks the Truth and wants the full story, should read his books." — JOHN BULL BADER

✯ "I love all of Colonel Seabrook's books!" — DEBBIE SIDLE

✯ "Lochlainn Seabrook is well educated and versed in what he writes and I'm impressed with the delivery." — THOMAS L. WHITE

✯ "Lochlainn Seabrook is the author of great works of scholarship." — JOHN B. (last name withheld)

✯ "Thank you Lochlainn Seabrook for your wonderful books! You are the real deal! You are an amazing author and I love your books!!" — SOPHIA MEOW CELLIST

✯ "I really enjoy Mr. Seabrook's books! His knowledge is beyond belief!" — SANDRA FISH

✯ "Love Lochlainn Seabrook. Awesome!!" — ROBIN HENDERSON ARISTIDES

✯ "Kudos to Lochlainn Seabrook who is a very good and informative professional truthful historian. We need more like him!" — AMY VACHON

If you enjoyed this book you may be interested in some of Colonel Seabrook's other popular titles:

☛ MANMADE: MALE INVENTORS WHO CREATED THE MODERN WORLD
☛ EVERYTHING YOU WERE TAUGHT ABOUT THE CIVIL WAR IS WRONG, ASK A SOUTHERNER!
☛ THE GREATEST JESUS MYSTERY OF ALL TIME: WHERE WAS CHRIST BETWEEN THE AGES OF 12 AND 30?
☛ THE CONCISE BOOK OF OWLS: A GUIDE TO NATURE'S MOST MYSTERIOUS BIRDS
☛ VITAMIN D: THE MIRACLE TREATMENT FOR NEARLY EVERY DISEASE AND HEALTH ISSUE
☛ UFOS AND ALIENS: THE COMPLETE GUIDEBOOK

Available from Sea Raven Press and wherever fine books are sold

9 781955 351560